SCAC

I0904192

FEB 2013

Consciousness

Consciousness

Confessions of a Romantic Reductionist

Christof Koch

The MIT Press
Cambridge, Massachusetts
London, England

© 2012 Massachusetts Institute of Technology

All rights reserved. No part of this book may be reproduced in any form by any electronic or mechanical means (including photocopying, recording, or information storage and retrieval) without permission in writing from the publisher.

MIT Press books may be purchased at special quantity discounts for business or sales promotional use. For information, please email special_sales@mitpress.mit.edu or write to Special Sales Department, The MIT Press, 55 Hayward Street, Cambridge, MA 02142.

This book was set in Syntax and Times Roman by Toppan Best-set Premedia Limited. Printed and bound in the United States of America.

Library of Congress Cataloging-in-Publication Data

Koch, Christof, 1956–
Consciousness : confessions of a romantic reductionist / Christof Koch.
  p. cm.
Includes bibliographical references (p.   ) and index.
ISBN 978-0-262-01749-7 (hardcover : alk. paper)
1. Consciousness. 2. Mind and body. 3. Free will and determinism. 4. Life. I. Title.
B808.9.K63   2012
153—dc23
2011040572

10 9 8 7 6 5 4 3 2

To Hannele

Contents

Preface   ix
Acknowledgments   xi

Chapter 1: In which I introduce the ancient mind–body problem, explain why I am on a quest to use reason and empirical inquiry to solve it, acquaint you with Francis Crick, explain how he relates to this quest, make a confession, and end on a sad note   1

Chapter 2: In which I write about the wellsprings of my inner conflict between religion and reason, why I grew up wanting to be a scientist, why I wear a lapel pin of Professor Calculus, and how I acquired a second mentor late in life   11

Chapter 3: In which I explain why consciousness challenges the scientific view of the world, how consciousness can be investigated empirically with both feet firmly planted on the ground, why animals share consciousness with humans, and why self-consciousness is not as important as many people think it is   23

Chapter 4: In which you hear tales of scientist-magicians that make you look but not see, how they track the footprints of consciousness by peering into your skull, why you don't see with your eyes, and why attention and consciousness are not the same   41

Chapter 5: In which you learn from neurologists and neurosurgeons that some neurons care a great deal about celebrities, that cutting the cerebral cortex in two does not reduce consciousness by half, that color is leached from the world by the loss of a small cortical region, and that the destruction of a sugar cube–sized chunk of brain stem or thalamic tissue leaves you undead   59

Chapter 6: In which I defend two propositions that my younger self found
nonsense—you are unaware of most of the things that go on in your head,
and zombie agents control much of your life, even though you confidently
believe that you are in charge   75

Chapter 7: In which I throw caution to the wind, bring up free will, *Der Ring
des Nibelungen*, and what physics says about determinism, explain the
impoverished ability of your mind to choose, show that your will lags behind
your brain's decision, and that freedom is just another word for feeling   91

Chapter 8: In which I argue that consciousness is a fundamental property of
complex things, rhapsodize about integrated information theory, how it
explains many puzzling facts about consciousness and provides a blueprint
for building sentient machines   113

Chapter 9: In which I outline an electromagnetic gadget to measure
consciousness, describe efforts to harness the power of genetic engineering
to track consciousness in mice, and find myself building cortical
observatories   137

Chapter 10: In which I muse about final matters considered off-limits to
polite scientific discourse: to wit, the relationship between science and
religion, the existence of God, whether this God can intervene in the
universe, the death of my mentor, and my recent tribulations   149

Notes   167
References   173
Index   179

Preface

What you're holding in your hand is a slim exposition on the modern science of consciousness. Within the space of a few hours, you can inform yourself about where we natural scientists stand with respect to unraveling one of the central questions of our existence—namely, how do subjective feelings, how does consciousness, enter into the world? "Through your head" is the obvious answer. But that answer is not very helpful. What is it about the brain inside your head that makes you conscious of colors, of pain and pleasure, of the past and of the future, of yourself and of others? And does any brain suffice? What about the brain of a comatose patient, of a fetus, of a dog, a mouse, or a fly? What about the "brains" of computers? Can they ever be conscious? I'll tackle these questions and then some, including free will, a theory of consciousness, and the *bête noire* of my research: the extent to which quantum mechanics is relevant to understanding consciousness.

This book is not just about science, however. It is also a confession and a memoir. I am not only a dispassionate physicist and biologist but also a human being who enjoys but a few years to make sense of the riddle of existence. I learned over the past years how powerfully my unconscious inclinations, my beliefs, and my personal strengths and failings have influenced my life and the pursuit of my life's work. I experienced what the novelist Haruki Murakami described in a striking interview: "We have rooms in ourselves. Most of them we have not visited yet. Forgotten rooms. From time to time we can find the passage. We find strange things . . . old phonographs, pictures, books . . . they belong to us, but it is the first time we have found them." You'll learn about some of these lost chambers as they become relevant to the quest I'm on—to uncover the roots of consciousness.

Pasadena, California
May 2011

Acknowledgments

Writing, editing, and publishing a book requires the active cooperation of many people. Books are a testament to the better nature of human-kind—taking pleasure in working toward a distant goal, with the primary reward being the feeling of a job well done.

Blair Potter took my prose and edited it. She identified the three distinct strands in my writing, un- and then re-braided them. If the outcome reads like anything close to a monolithic text, it is thanks to her. John Murdzek and Katherine Almeida proofread everything with a fine-tooth comb and Sara Ball, Amy Chung-Yu Chou, and Kelly Overly provided some more editorial advice.

Volney Gay, professor of psychiatry and of religious studies at Vanderbilt University in Nashville, invited me to give three *Templeton Research Lectures* on "The Problem of Consciousness in Philosophy, Religion, and Science" in spring 2007. It is here that inception for this book took place. I thank the John Templeton Foundation for their generous financial support for these public lectures.

I would like to acknowledge the many students, friends, and colleagues who read installments of the book and identified its many infelicities and inconsistencies—Ralph Adolphs, Ned Block, Bruce Bridgeman, McKell Ronald Carter, Moran Cerf, David Chalmers, Michael Hawrylycz, Constanze Hipp, Fatma Imamoglu, Michael Koch, Gabriel Kreiman, Uri Maoz, Mili Milosavljevic, Leonard Mlodinow, Joel Padowitz, Anil Seth, Adam Shai, Giulio Tononi, and Gideon Yaffe. Heather Berlin suggested the title. Bruce Bridgeman, McKell Carter, and Giulio Tononi took the time carefully to go through the entire text and emend it. Thanks to their collective efforts, their veiled or direct critiques, the book reads more smoothly, with fewer knobs that protrude, annoy, or distract.

Thanks to the many institutions that provided me a quiet haven. Foremost the California Institute of Technology, my intellectual home for a quarter of a century. During these past years, the hardest of my life, Caltech and its people were my one stable rock. They helped me to cope in ways both small and large. Korea University in Seoul provided a refuge in the Far East, with scope to write, think, and reflect upon all the matters discussed in these pages. The Allen Institute for Brain Science in Seattle generously gave me the time to finish this book.

Research in my laboratory is supported by the National Science Foundation, the National Institutes of Health, the Office of Naval Research, the Defense Advanced Research Projects Agency, the G. Harold & Leila Y. Mathers Foundation, the Swartz Foundation, the Paul G. Allen Family Foundation, and the World Class University program of the National Research Foundation of Korea. I am grateful to all.

Chapter 1: In which I introduce the ancient mind–body problem, explain why I am on a quest to use reason and empirical inquiry to solve it, acquaint you with Francis Crick, explain how he relates to this quest, make a confession, and end on a sad note

I can't tell you what it really is, I can only tell you what it feels like.
—Eminem, "Love the Way You Lie" (2010)

It was an everyday occurrence that set my life on a new path. I had already taken an aspirin, but the toothache persisted. Lying in bed, I couldn't sleep because of the pounding in my lower molar.

Trying to distract myself from this painful sensation, I wondered how it came to hurt so much. I knew that an inflammation of the tooth pulp sends electrical activity up one of the branches of the trigeminal nerve, which ends in the brain stem. After passing through further relays, nerve cells in part of the neocortical gray matter just beneath the skull become active and discharge their electrical impulses. Such bioelectrical activity in this part of the brain goes hand in hand with the consciousness of pain, including its awful, aching feeling.

But wait. Something profoundly inexplicable just happened. How can activity in the brain trigger feelings? It's just squishy stuff. How can mere meat, as cyberpunk novels dismissively refer to the body, engender sentience? Putting it more generally, how can anything physical give rise to something nonphysical, to subjective states? Whether it is the pain that I experienced on a distant summer day on the Atlantic shore, the joy I feel upon seeing my children, or the taste of a sparkling Vouvray, all have the same problematic origin in agitations of nervous matter.

It is problematic because of the seemingly unbridgeable gap between the nervous system and its interior view, the sensations that it generates. On the one hand is the brain, the most complex object in the known universe, a material thing subject to the laws of physics. On the other hand is the world of awareness, of the sights and sounds of life, of fear and anger, of lust, love, and ennui.

These two worlds are closely related—as a stroke or a strong blow to the head demonstrates dramatically. Oscar Wilde expressed it poetically, "It is in the brain that the poppy is red, that the apple is odorous, that the skylark sings." But exactly how does this transformation happen? How does the brain construct conscious experience? Through its shape, its size, its activity, its complexity?

Consciousness does not appear in the equations that make up the foundations of physics, nor in chemistry's periodic table, nor in the endless ATGC molecular sequences of our genes. Yet both of us—I, the author of these lines, and you, the reader—are sentient. That is the universe in which we find ourselves, a universe in which particular vibrations of highly organized matter trigger conscious feelings. It seems as magical as rubbing a brass lamp and having a djinn emerge who grants three wishes.

I am a nerd. As a kid, I built my own home computer to implement the Boolean laws of logic. I would lie awake in bed, designing elaborate tunnel-boring machines in my mind. So, it was natural for me to wonder during my toothache whether a computer could experience pain. Suppose that I coupled a temperature sensor to my laptop and programmed it in such a manner that if the room became too hot, the word "pain" would appear on its screen in big red letters. But would "pain" feel like anything to my Mac? I'm willing to grant many things to any Apple product, especially coolness, but not sentience.

But why not? Is it because my laptop operates on different physical principles? Instead of positively and negatively charged sodium, potassium, calcium, and chloride ions sloshing into and out of nerve cells, electrons flow onto the gates of transistors, causing them to switch. Is that the critical difference? I don't think so, for it seems to me that, ultimately, it must be the functional relationships of the different parts of the brain to each other that matter. And those can be mimicked, at least in principle, on a computer. Is it because people are organic, made out of bones, muscles, and nerves, whereas computers are synthetic, built out of titanium, copper wire, and silicon? That doesn't appear critical, either. So perhaps it is because humans evolved by chance and necessity, whereas machines were explicitly designed? The evolutionary history of animals is radically different from that of digital machines, a difference that is reflected in their distinct architecture. But I don't see how that affects whether one or the other is conscious. It has got to be the physical state of the system right now that makes a difference, not how it got to be the way it is.

What is the difference that makes a difference?

In philosophy, the difficulty of explaining why somebody can feel anything is often referred to as the *Hard Problem*. The term was coined by the philosopher David Chalmers. He made his reputation in the early 1990s by a closely argued chain of reasoning, leading him to conclude that conscious experience does not follow from the physical laws that rule the universe. These laws are equally compatible with a world without consciousness or with a different form of consciousness. There will never be a reductionist, mechanistic account of how the objective world is linked to the subjective one. The term Hard Problem, with its capital H, as in "Impossibly Hard," subsequently went viral. Nobody disputes that the physical and the phenomenal worlds are closely linked in billions of people every day of their lives, but why this should be so is the puzzle.

Dave taught me an important lesson about philosophers. I had invited him to speak in front of a neurobiological and engineering audience. Afterward, over a bottle of wine, I was astounded when he insisted that no empirical fact, no discovery in biology or conceptual advance in mathematics, could dissuade him of this unbridgeable gap between the two worlds. The Hard Problem was not amenable to any such advances. I was aghast. How could mere words, without the benefit of either a mathematical or a physico-empirical framework, establish anything with that degree of certainty? To me, he had a powerful argument, but certainly no proof.

Since then, I have encountered numerous philosophers who are utterly convinced of the truth of their ideas. Such confidence in one's own ideas—without being fazed by the myriad competing ideas of others, who can't all be right—is rare among natural scientists. Because of our constant experimental confrontation with messy Mother Nature, who forces us to modify our most brilliant and aesthetically pleasing theories, we've learned the hard way to not place too much trust in any one idea until it is established beyond a reasonable doubt.

Still, at some unconscious level, those arguments had an effect on me. They suggested that in seeking to understand the phenomenal world, science had finally met its match, that consciousness was resistant to rational explanation, immune to scientific analysis, beyond the ken of empirical validation. This was the entry point for religion. Religion has an intuitive, plausible explanation for the phenomenon of mind: We are conscious because we have an immaterial soul, our true, inner self. The soul is part and parcel of a transcendental reality, above and beyond the categories of space and time and causality. This soul strives to be united with God at the end of time. These were the traditional answers that I, growing up in a devout Roman Catholic family, was raised to believe in.

Religion and science are two modes of understanding the world, its origin, and its meaning. Historically, they have opposed each other. Ever since the beginning of the Age of Enlightenment, religion in the West has been in retreat, losing one battle after another. One setback was the Copernican revolution, which removed Earth from the center of the universe to the distal reaches of a galaxy containing a hundred billion stars. But the worst blow was delivered by Darwin's theory of evolution by natural selection. It removed humans from their God-given dominion over Earth and replaced the epic story of *Genesis* with a tale stretching across the eons, full of sound and fury, signifying nothing. Evolution dethroned humans from their exalted position; we are but one species among innumerable others. In the molecular signature of our genes we can trace our descent from the primate lineage and, in the remoteness of deep time, from pond scum.

As a consequence, much religious doctrine is incompatible with the modern world view. This mismatch should not be surprising, as the myths and doctrines that underpin the great monotheistic religions were formed at a time when little was known about the size, age, and evolution of Earth or the organisms populating it.

Many people believe that science leaches meaning out of human actions, hopes, and dreams, leaving desolation and emptiness in their place. The pioneering molecular biologist Jacques Monod expressed this forlorn sentiment chillingly:

Man must at last wake up out of his millenary dream and discover his total solitude, his fundamental isolation. He must realize that, like a gypsy, he lives on the boundary of an alien world; a world that is deaf to his music, and as indifferent to his hopes as it is to his sufferings and his crimes.

During my college days, this epigram, together with equally icy fragments by Friedrich Nietzsche and others, decorated the walls of my dormitory room. Eventually, I rebelled against this expression of existential, cosmic indifference.

At this point I have a confession to make. With perfect hindsight, I now realize that what drew me to studying consciousness was a compelling and entirely subterranean desire to justify my instinctual belief that life is meaningful. I thought that science could not explain how feelings came into the world. By giving the study of consciousness my all and failing in this endeavor, I was going to demonstrate to my own satisfaction that science is inadequate to the task of fully understanding the nature of the mind–body divide, that it cannot explain the essential mystery at the heart of phenomenal existence, and that Monod's desolate sentiments were misguided. In the end, this is not how it turned out. And so my toothache set me on a course to explore the seas of consciousness, with the Hard Problem as my lodestar.

I started studying the mind–body problem with Francis Crick, the physical chemist who, with James Watson, discovered the double-helical structure of DNA, the molecule of heredity, in 1953. This singular event, which ushered in the era of molecular biology, is the best documented and most celebrated example of a revolutionary scientific discovery. It was lauded with a Nobel Prize in 1962.

As recounted in *The Eighth Day of Creation*, Horace Freeland Judson's brilliant history of molecular biology, Francis subsequently established himself as the field's chief intellect. It was to him and his theoretical ideas that others looked for guidance in the exuberant and giddy race toward deciphering the universal code of life. When that goal was achieved, Francis's interest shifted from molecular biology to neurobiology. In 1976, at the age of sixty, he plunged into this new field while simultaneously moving from Cambridge in the Old World to California in the New World.

Over sixteen years, Francis and I wrote two dozen scientific papers and essays together. All of them focused on the anatomy and physiology of the primate brain and their link to consciousness. When we started this labor of love in the late 1980s, writing about consciousness was taken as a sign of cognitive decline. Retired Nobel laureates could do it, as could mystics and philosophers, but not serious academics in the natural sciences. Betraying an interest in the mind–body problem beyond that of a hobby was ill-advised for a young professor, particularly one who had not yet attained tenure. Consciousness was a fringe subject: graduate students, always finely attuned to the mores and attitudes of their elders, rolled their eyes and smiled indulgently when the subject came up.

But those attitudes changed. Together with a handful of colleagues— Bernie Baars, Ned Block, David Chalmers, Jean-Pierre Changeux, Stanislas Dehaene, Gerald Edelman, Steven Laureys, Geraint Rees, John Searle, Wolf Singer, and Giulio Tononi, to mention a few—we gave birth to a science of consciousness. Though still inchoate, that new science represents a true paradigm shift and a consensus that consciousness is a legitimate topic of scientific investigations.

The midwife attending its birth was the fortuitous development of brain-imaging techniques, allowing the human brain to be safely and routinely visualized in action. Those techniques have had an electrifying effect on popular culture: Magnetic resonance imaging (MRI) images of the brain, with their telltale hot spots, are iconic. They can be found on the covers of magazines, on T-shirts, and in movies.

Studying the biological basis of awareness has turned into a mainstream, legitimate subject of inquiry.

For the past two-and-a-half decades, I have mentored a group of two dozen students, postdoctoral fellows, and staff at the California Institute of Technology (Caltech) who are focused on such research. I've worked with physicists, biologists, psychologists, psychiatrists, anesthesiologists, neurosurgeons, engineers, and philosophers. I have participated in countless psychology tests. I have had my brain zapped with pulses of strong magnetic fields and with weak electric currents, stuck my head into MRI scanners to see what is inside my cranium, and had my brain waves recorded while I slept.

In this book, I highlight stories from the front lines of modern research into the neurobiology of consciousness. Just as light presupposes its absence, darkness, so consciousness presupposes the unconscious. As Sigmund Freud, Pierre Janet, and others realized in the late nineteenth century, much of what goes on in our head is inaccessible to our mind—is not conscious. Indeed, when we introspect, we routinely deceive ourselves, because we only tap into a minute fraction of what is going on in our head. This deception is why so much of philosophy about the self, the will, and other aspects of our mind has been barren for more than two thousand years. Yet, as I shall describe, the unconscious can profoundly influence our behavior. I also dwell on the related problem of free will, the feeling of having initiated an action, and on how physics, psychology, and neurosurgery are untangling this metaphysical knot. Without much fanfare, discoveries in these fields have solved a key aspect of the free will problem.

Finally, I describe a plausible quantitative theory of consciousness that explains why certain types of highly organized matter, in particular brains, can be conscious. The theory of *integrated information*, developed by the neuroscientist and psychiatrist Giulio Tononi, starts with two basic axioms and proceeds to account for the phenomenal in the world. It is not mere speculative philosophy, but leads to concrete neurobiological insights, to the construction of a consciousness-meter that can assess the extent of awareness in animals, babies, sleepers, patients, and others who can't talk about their experiences. The theory has profound consequences that bear some resemblance to the prophetic ideas of Pierre Teilhard de Chardin (more of him later).

Discoveries in astronomy and cosmology reveal that the laws of physics are conducive to the formation of stable, heavy elements beyond hydrogen and helium. These laws are amazingly fine-tuned and require a precise balancing of the four fundamental physical forces. Otherwise, our universe would never have gotten to the stage where hydrogen and helium assembled into huge flaming masses—long-lived stars that provide an endless stream of energy to the rocky planets orbiting them. The stuff out of which these planets and their skin of soil, rock, and air is made—silicon, oxygen, and so on—was created inside the nuclear furnaces of the first generation of stars and dispersed into the surrounding space during their explosive death throes. We are, quite literally, star dust. This dynamic universe is governed by the second law of thermodynamics: The entropy of any closed system never decreases; or, in other words, the universe is unfolding to be maximally disordered and uniform. But this does not preclude the formation of stable islands of order that feed upon the surrounding ocean of free energy. The relentless operation of this law created the statistical certainty that on some such isles in the cosmos, long-chained, complex molecules would eventually arise.

Once this crucial step occurred, the next one was likely to happen as well: genesis—the creation of life in a cave or pond on the primordial Earth and elsewhere, under alien skies. The ever-increasing complexity of organisms, evident in the fossil record, is a consequence of the unrelenting competition for survival that propels evolution.

It was accompanied by the emergence of nervous systems and the first inkling of sentience. The continuing *complexification* of brains, to use Teilhard de Chardin's term, enhanced consciousness until self-consciousness emerged: awareness reflecting upon itself. This recursive process started millions of years ago in some of the more highly developed mammals. In *Homo sapiens*, it has achieved its temporary pinnacle.

But complexification does not stop with individual self-awareness. It is ongoing and, indeed, speeding up. In today's technologically sophisticated and intertwined societies, complexification is taking on a supraindividual, continent-spanning character. With the instant, world-wide communication afforded by cell phones, e-mail, and social network-ing, I foresee a time when humanity's teeming billions and their computers will be interconnected in a vast matrix—a planetary *Übermind*. Provided mankind avoids Nightfall—a thermonuclear Armageddon or a complete environmental meltdown—there is no reason why this web of hypertro-phied consciousness cannot spread to the planets and, ultimately, beyond the stellar night to the galaxy at large.

Now you know why the neuropsychologist Marcel Kinsbourne calls me a *romantic reductionist*: reductionist, because I seek quantitative explanations for consciousness in the ceaseless and ever-varied activity of billions of tiny nerve cells, each with their tens of thousands of syn-apses; romantic, because of my insistence that the universe has contrails of meaning that can be deciphered in the sky above us and deep within us. Meaning in the sweep of its cosmic evolution, not necessarily in the lives of the individual organisms within it. There is a *Music of the Spheres*, and we can hear snatches of it, perhaps even a hint of the whole form of it, if we but listen closely.

The subtitle of this book contains the promissory word "confessions." In the evolution of the genre, from Saint Augustine, who invented it in the twilight years of the Roman Empire, to today's talk and reality shows, there has always been a whiff, if not a stench, of the exhibitionist, the self-serving, and the mendacious. I intend to stay clear of those mal-odorous corruptions. I also write in the face of a powerful professional edict against bringing in subjective, personal factors. This taboo is why scientific papers are penned in the desiccated third person: "It has been shown that. . . ." Anything to avoid the implication that research is done by flesh-and-blood creatures with less than pristine motivations and desires.

In the following pages, I'll tell you about myself insofar as my life is relevant to the questions: Why was I motivated—consciously or otherwise—to pursue certain problems? And, why did I adopt a particu-lar scientific stance? It is, after all, in the choice of what we work on that we reveal much about our inner drives and motives.

In the past few years, as the arc of my life has begun its inevitable decline, I've lost my way. Passions that I could not, or would not, control led to a profound crisis that forced me to confront my beliefs and my inner demons. Dante's opening stanza in his *Inferno* describes it perfectly:

Midway in the journey of our life
I came to myself in a dark wood,
for the straight way was lost.

But before I engage too much in such late-night matters, let me tell you a bit about my early life that is relevant to my science and to the way I view the brain.

Chapter 2: In which I write about the wellsprings of my inner conflict between religion and reason, why I grew up wanting to be a scientist, why I wear a lapel pin of Professor Calculus, and how I acquired a second mentor late in life

Consider ye the seed from which ye sprang;
Ye were not made to live like unto brutes,
But for pursuit of virtue and of knowledge.
So eager did I render my companions,
With this brief exhortation, for the voyage,
That then I hardly could have held them back.
And having turned our stern unto the morning,
We of the oars made wings for our mad flight,
Evermore gaining on the larboard side.
—Dante Alighieri, *Inferno* (1531)

I grew up happy, obsessed with knowledge, structure, and order. My two brothers and I were raised by our parents in the best liberal Catholic tradition, in which science—including evolution by natural selection—was by and large accepted as explaining the material world. I was an altar boy, reciting prayers in Latin and listening to Gregorian chants and the masses, passions, and requiems of Orlande de Lassus, Bach, Vivaldi, Haydn, Mozart, Brahms, and Bruckner. During summer vacations, our family traveled to sightsee a countless stream of museums, castles, and baroque and rococo churches. My parents and older brother gazed in admiration at ceilings, stained-glass windows, statues, and frescos depicting religious imagery while my mother read aloud, for the benefit of us all, the detailed history of each object. Although I found this forced diet of art excruciatingly boring, and can't quite suppress a shudder even today when I see the three-volume art guide on my mother's bookshelf, I fell in love with the magical intonations of centuries-old Roman prayers and the sacred and profane music of these composers.

Mother Church was an erudite, globe-spanning, culturally fecund, and morally unassailable institution with an unbroken lineage extending across two millennia to Rome and Jerusalem. Its catechism offered a time-honored and reassuring account of life that made sense to me. So strong was the comfort religion provided that I passed it on. My wife and I raised our children in the faith, baptizing them, saying grace before meals, attending church on Sundays, and taking them through the rite of First Communion.

Yet over the years, I began to reject more and more of the church's teachings. The traditional answers I was given were incompatible with a scientific world view. I was taught one set of values by my parents and by my Jesuit and Oblate teachers, but I heard the beat of a different drummer in books, lectures, and the laboratory. This tension left me with a split view of reality. Outside of mass, I didn't give much thought to questions of sin, sacrifice, salvation, and the hereafter. I reasoned about the world, the people in it, and myself in entirely natural terms. These two frameworks, one divine and one secular, one for Sunday and one for the rest of the week, did not intersect. The church provided meaning by placing my puny life in the context of the vastness of God's creation and his Son's sacrifice for humankind. Science explained facts about the actual universe I found myself in and how it came to be.

Harboring two distinct accounts, one for the supralunar and one for the sublunar world, to use beautiful Aristotelian imagery, is not a serious intellectual stance. I had to resolve the conflict between these two types of explanations. The resultant clash was my constant companion for decades. Yet I always knew that there is but a single reality out there, and science is getting increasingly better at describing it. Humanity is not condemned to wander forever in an epistemological fog, knowing only the surface appearance of things but never their true nature. We can see something; and the longer we gaze, the better we comprehend.

It is only in recent years that I have managed to resolve this conflict. I slowly but surely lost my faith in a personal God. I stopped believing that somebody watches over me, intervenes on my behalf in the world, and will resurrect my soul beyond history, in the eschaton. I lost my childhood faith, yet I've never lost my abiding faith that everything is as it should be! I feel deep in my bones that the universe has meaning that we can realize.

A Carefree Childhood as a Budding Scientist

My father had studied law, joined the German foreign service, and became a diplomat. My mother was a doctor and worked for a few years in a hospital. She gave up her career for the sake of her husband, channeling her considerable ambitions into us children.

I was born in 1956 in Kansas City, Missouri, one year after my brother Michael. Today, you can't tell my Midwestern origin, as I retain a fairly strong German accent. We left two years later and started a peripatetic existence, staying four years in Amsterdam, where my younger brother, Andreas, was born. Subsequently, our family lived in Bonn, then the capital of West Germany. After elementary public school and two years at a Jesuit *Gymnasium*, it was time to move back across the Atlantic, to Ottawa. I learned English in a school run by a Catholic religious order. But not for long, for three years later it was to be Rabat, Morocco. I enrolled in the French school in that North African country, the thoroughly secular *Lycée Descartes* (which may explain my enduring fondness for that particular philosopher). Despite the constant changes of place, school, and friends, and the need to master a third language, I did well, graduating in 1974 with a baccalaureate in mathematics and sciences.

I am fortunate in that I knew from an early age what I wanted to be when I grew up. As a child I dreamt of being a naturalist and zoo director, studying animal behavior on the Serengeti. Around the onset of puberty, my interests shifted to physics and mathematics. I consumed a steady diet of popular books about space travel, quantum mechanics, and cosmology. I loved the paradoxes of relativistic travel, of falling past the time horizon into a black hole, space elevators, and so on. I have fond memories of reading the whimsical *Mr. Tompkins in Wonderland* by George Gamow, in which the hero explores a surreal world where the speed of light can be reached by a cyclist. Or *Mr. Tompkins Explores the Atom*, in which Planck's constant, the number that describes the size of quanta of action, is so large that billiard balls show quantum behavior. These books molded my teenage mind. Each time I bought a science paperback with my weekly allowance, I lovingly inscribed my name in it and cherished it, carrying it around everywhere, reading it whenever I could.

My parents furthered my scientific curiosity by giving Michael and me experimental designer sets from the German brand *Kosmos*. These were refined toys, to teach physics, chemistry, electronics, or astronomy via a series of do-it-yourself experiments. One set started with the basic laws of electricity, went from there to the assembly of an electromagnetic relay and an induction motor, and ended with my building an AM and then an FM radio receiver. I spent hours and hours messing around with the electronics, the kind of hardware hacking that today's kids rarely experience. Another set taught the principles of inorganic chemistry. I used my freshly acquired know-how to mix black powder. When I constructed a bazooka, and the metal rod that was supposed to guide the rocket melted (the propellant didn't ignite fast enough), my father intervened, aborting my short-lived career as a weapons designer. In the process, he probably saved my limbs and eyes.

My father bought us a 5-inch reflector telescope, an awesome instrument. I vividly remember one night on the rooftop of our house in Rabat, when Michael and I calculated the position of Uranus on a star map to the background strains of Wagner's *The Flying Dutchman*. What elation I felt when, pointing the scope to the estimated azimuth and elevation in the sky, the shimmering planet gently drifted into view. What a terrific confirmation of order in the universe!

During my North African sojourn, I became permanently enchanted by the adventures of Tintin, the Belgian boy whose official job is reporter but who is really an explorer, detective, and all-round hero; his white Fox Terrier, Snowy (Milou in the original French); his boisterous friend Captain Haddock; and the mad scientist, resident genius, brilliant but absent-minded, and nearly deaf Professor Calculus. These were the first cartoon characters I encountered, because my parents frowned upon comics as being either too childish or too crass. I have given every one of the twenty-four Tintin books to my children, who love them too, with no apparent ill effect. Tintin posters even grace the hallway of my house. Professor Calculus is the archetype of the otherworldly scholar who understands the secrets that hold the universe together, yet who is a klutz when dealing with the everyday world. He had such a formative effect on my young mind that I have worn a pin with his figure on the lapel of my jacket since the day in April 1987 when I gave my inaugural lecture as a professor.

Growing up in different countries, attending different schools, and learning different languages allowed me, more so than my less mobile friends, to see beyond the peculiarities and distinctiveness of any one culture and appreciate the underlying universal traits. This was one of many formative reasons that made me, by the time I left home, want to be a physicist.

In 1974, I enrolled at the University of Tübingen, in southwest Germany. Tübingen is a quaint, small academic town built around a castle, much like its better-known rival, Heidelberg. At university I experienced the fellowship of a band of brothers in a fencing fraternity. If you are not steeped in Teutonic academic traditions, think of the Boy Scouts transposed to a romantic, 500-year-old university to get some idea of what I mean. I also became acquainted with and indulged in, sometimes excessively so, the pleasures and perils of alcohol, women, dancing, Friedrich Nietzsche, and Richard Wagner. I spent the first Christmas away from home—a friend and I sequestered ourselves in a remote village, enraptured by our readings of *Thus Spoke Zarathustra* and the lyrics and music of *Tristan and Isolde* and *Der Ring des Nibelungen*. I was young, immature, and nerdy, and I needed to take this voyage of self-discovery through the noisy and glorious confusion of life.

In 1979, I graduated from the University of Tübingen with a master's degree in physics. On the way, I had acquired a minor in philosophy, which led me to idealism, the form of monism which teaches that the universe is but a manifestation of the mind.

By then it had dawned on me that I didn't quite have the mathematical skills necessary to be a world-class cosmologist. Fortunately, around that time I became enthralled by computers. What attracted me was their promise of creating a self-contained virtual world under my complete control. Within their simplified environment, all events follow the rules—the algorithm—laid down by the programmer. Any deviation can always be traced back to faulty reasoning or incomplete assumptions. If a program didn't work, you had nobody to blame but yourself and yourself alone. I wrote programs, initially on punched cards submitted to the university's centralized computer, in Algol and assembler language for astrophysicists and nuclear physicists.

Studying the Biophysics of Nerve Cells

I also became utterly fascinated with the notion that the brain is a kind of computer, processing information. This obsession was triggered by an inspirational book, *On the Texture of Brains: Neuroanatomy for the Cybernetically Minded*, written by the German-Italian anatomist Valentino Braitenberg. Valentino has a larger-than-life personality, living proof that one can be a great scientist, an esthete, a musician, a bon vivant, and a *mensch*, all at the same time.

Valentino was a director at the Max Planck Institute for Biological Cybernetics in Tübingen. Through him, I found a job writing code for the Italian physicist Tomaso Poggio at the institute. Tommy, as everybody calls him, is one of the world's great theoreticians of information processing. He invented the first functional formula for extracting stereo depth from two disparate views of the same scene. Under his guidance,

I did my thesis work, modeling on a computer how the excitatory and inhibitory synapses placed on a single nerve cell interact with each other.

Let me briefly digress to explain a couple of basic concepts that occur throughout this book. Like all organs, the nervous system is made out of billions of networked cells, the most important of which are neurons. Just like there are kidney cells that are quite distinct from blood or heart cells, so there are different types of neurons, maybe as many as a thousand. The most important distinction between them is whether they excite or inhibit the neurons they are connected to. Neurons are highly diverse and sophisticated processors that collect, process, and broadcast data via synapses, or contact points with other nerve cells. They receive input via their finely branched dendrites, which are studded with thousands of synapses. Each synapse briefly increases or decreases the electrical conductance of the membrane. The resultant electrical activity is translated, via sophisticated, membrane-bound machinery in the dendrites and the cell body, into one or more all-or-none pulses, the fabled action potentials, or spikes. Each of these pulses is about one-tenth of a volt in amplitude and lasts less than one-thousandth of a second. These pulses are sent along the neuron's output wire, the axon, which connects to other neurons via synapses. (Some specialized neurons send their output to muscles.) Thus, the circle closes. Neurons talk to other neurons via synapses. This is the habitat of consciousness.

The power of the nervous system is found not in the snail-like speed of its components, but in its massive parallel communication and computation capabilities: its ability to link very large and highly heterogeneous coalitions of neurons over large distances in very specific synaptic patterns. As I would demonstrate thirty years later, it is out of these patterns that our thoughts arise. Synapses are analogous to transistors. Our nervous system has perhaps 1,000 trillion synapses linking about 86 billion neurons.

Under Tommy's guidance, I solved the differential equations that describe how the electrical charge inside and outside of the membrane surrounding a nerve cell is transformed by the branching pattern of its dendrites and the architecture of its synapses. Today, such modeling is routine and well respected, but back then biologists were baffled by the use of physics to describe events in the brain. At the first national meeting where I presented my research in the form of a poster to other scientists, I was quarantined to the back of the conference hall. Only two visitors came by, one of whom was looking for the bathroom but was polite enough to stay and talk to me. I got drunk that night, wondering whether I had chosen the right field. Despite such setbacks, I graduated in 1982 with a Ph.D. in biophysics.

During my years as a doctoral candidate, I fell in love with and married Edith Herbst. Edith is a nurse, born and raised in Tübingen. While pregnant with our son, Alexander, she typed my thesis into the institute's mainframe (with a 128-kilobyte core memory!). When my thesis advisor, my *Doktorvater* as we say endearingly in German, became a professor at the Massachusetts Institute of Technology (MIT), we followed him to Cambridge. Twenty-five years old, I was starting off in a foreign country as a postdoctoral fellow.

MIT was an intellectual blast. I stayed for four years in the Department of Psychology and the Artificial Intelligence Laboratory. I was free to pursue science, pure and simple. Remaining so long with the same advisor is irregular, but it worked out well for my career. Tommy and I continue to interact today, a testament to the longevity of the link between doctor-father and son.

Caltech, Teaching, Research, and the Brain Viewed by a Physicist

In fall 1986, I moved farther west with my family—enlarged by a daughter Gabriele—arriving at the California Institute of Technology as an assistant professor of biology and engineering. Caltech, one of America's most selective, hard-core science and engineering universities, lies in Pasadena, a suburb of Los Angeles. It is crisscrossed by broad avenues lined with palm, orange, and oak trees, nestled at the foot of the San Gabriel Mountains. I was immensely proud of joining Caltech's faculty.

Caltech is a nimble, private university—about 280 professors and 2,000 undergraduate and graduate students—dedicated to educating the best and brightest in logic, mathematics, and how to reason about the natural world. Caltech and its people embody everything that is great and noble about universities—institutions that have been in existence for eight hundred years. It is an ivory tower in the best sense of the term, affording ample freedom and resources to pursue the nature of consciousness and the brain.

The first thing anybody asks when they find out that I'm a professor is, "What do you teach?" In the eye of the public, a professor is fully defined by his or her traditional role as a teacher. I enjoy "professing" and teach a variety of classes. Instructing brilliant and motivated students who have no hesitation in pointing out errors or inconsistencies is an intellectual challenge of the highest order and emotionally rewarding. Time and again I've had insights while preparing my lectures or answering questions in class that illuminated a well-worn problem from an unexpected angle.

Yet the members of my tribe derive their self-esteem and sense of worth principally from their research. Where we are in the tribal hierarchy is determined by how successful our investigations are. Research is what drives us and is our greatest source of pleasure. The yardstick used to gauge success is the number and quality of our publications in high-profile, peer-reviewed, and hypercompetitive scientific journals.

The greater the impact our discoveries have in this rarefied world, the bigger our reputation. Teaching plays only a minor role in the field's collective self-image. We professors spend the bulk of our time on investigations—thinking, reasoning and theorizing, computing and programming, talking thru ideas with colleagues and co-workers, reading the commodious literature, contributing ourselves to it, speaking at seminars and conferences, generating the countless grant proposals that feed the research machinery and keep it lubricated, and, of course, supervising and mentoring students and postdoctoral fellows who design, fabricate, measure, shake, stir, image, scan, record, analyze, program, debug, and compute. I'm the chieftain of a band of about two dozen such researchers.

Besides characterizing selective visual attention and visual consciousness—more on that later—we continue to investigate the biophysics of neurons. The brain is a highly evolved organ, yet it is also a physical system that obeys ironclad laws of conservation of energy and of electrical charge. Gauss's and Ohm's laws regulate the distribution of charges inside and outside of nerve cells and their associated electric fields. All the synaptic and spiking processes described above contribute to the electrical potential that is picked up by electrodes stuck into the brain's gray matter. If tens of thousands of neurons and their millions of synapses are active, their contributions add up to something called the local field potential. The distant echo of this electrical activity is visible in the never-ceasing peaks and troughs recorded outside the skull by an electroencephalograph (EEG). The local field potential, in turn, feeds back onto individual neurons. We are now learning that this feedback forces neurons to synchronize their activity.

This reciprocal interaction between the local action of neurons and the global field they collectively produce and are enveloped by is very different from electronic silicon circuits, whose designers lay out wires, transistors, and capacitances to avoid interference with each other, keeping "parasitic" cross-talk at a minimum. I'm very interested in the brain's electrical field, how much information is carried by this field, and its possible role in consciousness.

What is striking to a physicist studying the brain and the mind is the absence of any conservation laws: Synapses, action potentials, neurons, attention, memory, and consciousness are not conserved in any meaningful sense. Instead, what biology and psychology do have in exuberant abundance are empirical observations—facts. There is no unifying theory, with the singular exception of Darwin's theory of evolution by natural selection—and although it is an exceedingly powerful explanatory framework, evolutionary theory is open-ended, and not predictive. Instead, the life sciences have lots of heuristics, semi-exact rules, that capture and quantify phenomena at one particular organismal scale—such as the biophysical modeling that I worked on for my thesis—without aspiring to universality. That makes research in these fields quite different from physics.

Once More unto the Breach of Consciousness

When I started off in California, I became reacquainted with Francis Crick. I first encountered him lying under an apple tree in an orchard outside of Tübingen in the summer of 1980. Francis had come to town to talk—his favorite activity—with Tommy about our ongoing dendritic and synaptic modeling work.

Four years later and on another continent, Francis had invited me and Shimon Ullman, a computer scientist at MIT's Artificial Intelligence Laboratory, to the Salk Institute for five days. Francis wanted to know everything about a model of selective visual attention that Shimon and I had just published. Why this particular wiring scheme? How many neurons were involved? What was their average firing rate? How many synapses did they form? What was their time constant? To what part of the thalamus did their axons project? Could that account for the celerity of the behavioral response? This went on and on from breakfast until late afternoon. A break was followed by dinner and more conversations focused on the brain. No idle chatter for Francis. It left me breathless. I admired how his wife, Odile, dealt with this intensity for decades.

A couple of years later, Francis and I started to collaborate, with daily phone calls, letters, e-mails, and monthly extended stays at his hillside home in La Jolla, a two-hour drive south of Caltech. The focus of our work was consciousness. Although generations of philosophers and scholars had tried in vain to solve the mind–body problem, we believed that a fresh perspective from the vantage point of the neurosciences might help unravel this Gordian knot. As a theoretician, Francis's methods of inquiry were quiet thinking, daily reading of the relevant literature—he could absorb prodigious amounts of it—and the Socratic dialogue. He had an unquenchable thirst for details, numbers, and facts. He would ceaselessly put hypotheses together to explain something, then reject most of them himself. In the morning, he usually bombarded me with some bold new hypothesis that had come to him in the middle of the night, when he couldn't sleep. I slept much more soundly and, therefore, lacked such nocturnal insights.

In a lifetime of teaching, working, and debating with some of the smartest people on the planet, I've encountered brilliance and high achievement, but rarely true genius. Francis was an intellectual giant, with the clearest and deepest mind I have ever met. He could take the same information as anybody else, read the same papers, yet come up with a totally novel question or inference. The neurologist and author Oliver Sacks, a good friend of us both, recollects that the experience of meeting Francis was "a little like sitting next to an intellectual nuclear reactor . . . . I never had a feeling of such *incandescence*." It has been said that Arnold Schwarzenegger, in his heyday as Mr. Universe, had muscles in places where other people didn't even have places. This *bon mot*, transcribed to the rational mind, applied to Francis.

Equally remarkable was how approachable Francis was. No celebrity attitude for him. Like James Watson, I, too, have never seen Francis in a modest mood. But neither have I seen him in an arrogant mood. He was willing to talk to anybody, from lowly undergraduate student to fellow Nobel laureate, provided that the interlocutor brought him interesting facts and observations, a startling proposal, or some question he had never previously considered. It is true that he would quickly lose patience with people who spouted nonsense or didn't understand why their reasoning was wrong, but he was one of the most open-minded savants I have ever known.

Francis was a reductionist writ large. He fiercely opposed any explanation that smacked even remotely of religion or woolly-headed thinking, an expression he was fond of using. Yet neither my metaphysical sentiments nor our forty-year age difference prevented us from developing a deep and abiding mentor–student relationship. He relished the opportunity to endlessly bounce ideas off a younger man with plenty of energy, domain-specific knowledge, a love of speculation, and the temerity to sometimes disagree vigorously with him. I was very fortunate that he took a liking to me, essentially adopting me as his intellectual son.

Let me now define the problem of consciousness and describe the approach Francis and I took to explore its nature.

Chapter 3: In which I explain why consciousness challenges the scientific view of the world, how consciousness can be investigated empirically with both feet firmly planted on the ground, why animals share consciousness with humans, and why self-consciousness is not as important as many people think it is

How did evolution convert the water of biological tissue into the wine of consciousness?
—Colin McGinn, *The Mysterious Flame* (1999)

Without consciousness there is nothing. The only way you experience your body and the world of mountains and people, trees and dogs, stars and music is through your subjective experiences, thoughts, and memories. You act and move, see and hear, love and hate, remember the past and imagine the future. But ultimately, you only encounter the world in all of its manifestations via consciousness. And when consciousness ceases, this world ceases as well.

Many traditions view humans as having a mind (or psyche), a body, and a transcendental soul. Others reject this tripartite division in favor of a mind–body duality. The ancient Egyptians and Hebrews placed the psyche in the heart, whereas the Maya located it in the liver. We moderns know that the conscious mind is a product of the brain. To understand consciousness, we must understand the brain.

But there's the rub. How the brain converts bioelectrical activity into subjective states, how photons reflected off water are magically transformed into the percept of an iridescent aquamarine mountain tarn is a puzzle. The nature of the relationship between the nervous system and consciousness remains elusive and the subject of heated and interminable debates.

The seventeenth-century French physicist, mathematician, and philosopher René Descartes, in his *Discourse on the Method of Rightly Conducting the Reason, and Seeking Truth in the Sciences*, sought ultimate certainty. He reasoned that everything was open to doubt, including whether the outside world existed or whether he had a body. That he was experiencing something, though, even if the precise character of what he was experiencing was delusional, was a certainty. Descartes concluded that because he was conscious, he existed: *Je pense, donc je suis*, later translated as *cogito, ergo sum*, or *I think, therefore I am*. This statement brings home the fundamental importance of consciousness: It is not a rare state that comes to you only when you are contemplating your navel while sitting in the lotus position on a mountaintop, humming "Om" through your nostrils. Unless you are deeply asleep or in a coma, you are always conscious of something. Consciousness is the central fact of your life.

The singular point of view of the conscious, experiencing observer is called the *first-person perspective*. Explaining how a highly organized piece of matter can possess an interior perspective has daunted the scientific method, which in so many other areas has proved immensely fruitful.

Consider measurements made by NASA's Cosmic Background Explorer (COBE) satellite. In 1994, COBE became front-page news with its oval-shaped, blue-greenish photo of the full sky with smudges of yellow and red. The warm colors denoted tiny variations in the temperature of the cosmic background radiation, a relic of the Big Bang that produced our universe. Cosmologists can listen to the echoes of this titanic explosion of space itself and infer the shape of the early universe. The COBE data confirmed their expectations; that is, astronomy can make testable statements about an event that took place 13.7 billion years ago! Yet something as mundane as a toothache, right here and now, remains baffling.

Biologists can't yet specify the detailed molecular programs inside a single fertilized mammalian egg that turn it into the trillion cells making up the liver, muscle, brain, and other organs of a fully formed individual. But there is no doubt that the essential tools for such a feat are at hand. Molecular engineers, under the leadership of the scientist–entrepreneur Craig Venter, achieved a milestone in 2010 with the creation of a new species. They sequenced the genome (a single stretch of DNA one million letters long) of a bacterium, added identifying watermarks to it, synthesized the genes that made up the genome (using the four types of chemicals that make up DNA), and assembled them into one string that was then implanted into the cell body of a donor bacterium whose own DNA had been removed. The artificial genome successfully commanded the donor cell's protein-making machinery, and the new organism, dubbed *Mycoplasma mycoides JCVI-syn1.0*, proceeded to replicate generation after generation. Although the creation of a new bacterial species does not make a *golem*, it is nevertheless an amazing act, a watershed moment in history. There are no theoretical barriers to programming simple, multicellular plants and animals in this manner, only immense practical ones. But those will eventually be overcome. The ancient dream of the alchemist—the creation of life in the laboratory—is within reach.

In 2009, I attended Science Foo (SciFoo) Camp at the Googleplex in Mountain View, California. A few hundred übergeeks, technologists, scientists, space enthusiasts, journalists, and digerati assembled for a weekend of improvised seminars and exchanges. There was much talk about the future of artificial intelligence, with some arguing that the quest for true artificial intelligence, say at the level of a six-year-old child, had been abandoned. Yet nobody doubted that software constructs whose intelligence will rival, and ultimately exceed, ours will come about. Whereas computer scientists and programmers are probably several decades away from achieving human intelligence, there is, in principle, no difficulty in doing so. Nobody at SciFoo thought it inconceivable reaching this goal. The best way of doing so was debated, as was the question of how to address the problem of learning, whether artificial intelligence was going to be good or bad for society, and so on. But no one questioned that the goal was attainable.

There is nothing comparable for consciousness, no consensus regarding the likelihood of understanding its physical basis. John Tyndall, the Irish-born physicist who discovered why the sky is blue and that water vapor and carbon dioxide are the two biggest heat-trapping greenhouse gases in Earth's atmosphere, well articulated the difficulty of linking consciousness to the brain as far back as 1868:

The passage from the physics of the brain to the corresponding facts of consciousness is unthinkable as a result of mechanics. Granted that a definite thought, and a definite molecular action in the brain, occur simultaneously; we do not possess the intellectual organ, nor apparently any rudiment of the organ, which would enable us to pass, by a process of reasoning, from the one phenomenon to the other. They appear together, but we do not know why. Were our minds and senses so expanded, strengthened, and illuminated, as to enable us to see and feel the very molecules of the brain; were we capable of following all their motions, all their groupings, all their electric discharges, if such there be; and were we intimately acquainted with the corresponding states of thought and feeling, we should be as far as ever from the solution of the problem, "How are these physical processes connected with the facts of consciousness?" The chasm between the two classes of phenomena would still remain intellectually impassable. Let the consciousness for love, for example, be associated with a right-handed spiral motion of the molecules of the brain, and the consciousness of hate with a left-handed spiral motion. We should then know, when we love, that the motion is in one direction, and, when we hate, that the motion is in the other; but the "WHY?" would remain as unanswerable as before.

This challenge is what Chalmers refers to as the Hard Problem.

Neuroscientists peer at the nervous system through microscopes and magnetic scanners, they map the minutiae of its physical layout, they stain its neurons in all the colors of the rainbow, and they listen to the faint whispering of the neurons inside the brain of a monkey or a person looking at pictures or playing a video game. The latest technocraze, one likely to net Nobel Prizes for the pioneers who developed it, is *optogenetics*. This method targets specific groups of nerve cells deep inside an animal's brain that have been infected with modified viruses. The viruses cause the neurons to sprout photoreceptors that respond only to light of a specific wavelength. The neurons can then be turned on with brief pulses of blue light and turned off with equally brief pulses of yellow light. Playing the light organ of the brain! Optogenetics is hot stuff because it allows researchers to intercede deliberately at any point within the tightly woven networks of the brain, moving from observation to manipulation, from correlation to causation. Any group of neurons with a unique genetic barcode can be shut off or turned on with unparalleled precision. I shall return to this immensely promising technology in the penultimate chapter.

All of these techniques measure and perturb the nervous system from a *third-person perspective*. But how does nervous tissue acquire an interior, first-person point of view? Let's go back another 150 years, to the *Monadology* of Gottfried Leibniz. He was the German mathematician, scientist, and philosopher referred to as "the last man who knew everything" (he co-invented infinitesimal calculus and the modern binary number system). Leibniz wrote in 1714:

Moreover, we must confess that the perception, and what depends on it, is inexplicable in terms of mechanical reasons, that is, through shapes and motions. If we imagine that there is a machine whose structure makes it think, sense, and have perceptions, we could conceive it enlarged, keeping the same proportions, so that one could enter into it, as one enters into a mill. Assuming that, when inspecting its interior, we will only find parts that push one another, and we will never find anything to explain a perception.

Many scholars consider this gap between the mechanisms of the brain and the consciousness it exudes to be unbridgeable. If consciousness remains unaccounted for, science's explanatory domain is far more limited than its practitioners would like to believe and its propagandists shout from the rooftops. The inability to explain sentience within a quantitative, empirically accessible framework would be scandalous.

I do not share this defeatist attitude. Despite the hectoring of deconstructionist, "critical" scholars and sociologists, science remains humanity's most reliable, cumulative, and objective method for comprehending reality. It is far from fail-safe; it is beset with plenty of erroneous conclusions, setbacks, frauds, power struggles among its practitioners, and other human foibles. But it is far better than any alternative in its ability to understand, predict, and manipulate reality. Because science is so good at figuring out the world around us, it should also help us to explain the world within us.

Scholars don't know why our inner, mental world exists at all, let alone what it is made of. This persistent puzzle makes consciousness irksome to some and anathema to many of my colleagues. Yet the resistance of consciousness to a reductionist understanding delights much of the public. They denigrate reason and those who serve its call, for a complete elucidation of consciousness threatens long and dearly held beliefs about the soul, about human exceptionalism, about the primacy of the organic over the inorganic. Dostoyevsky's Grand Inquisitor understands this mind-set well: "There are three powers, three powers alone that are able to conquer and hold captive for ever the conscience of these impotent rebels for their happiness—those forces are miracle, mystery, and authority."

Qualia and the Natural World

At this point, I need to introduce *qualia*, a concept beloved by philosophers of mind. Qualia is the plural of *quale*. What it feels like to have a particular experience is the quale of that experience: The quale of the color red is what is common to such disparate percepts as seeing a red sunset, the red flag of China, arterial blood, a ruby gemstone, and Homer's wine-dark sea. The common denominator of all these subjective states is "redness." Qualia are the raw feelings, the elements that make up any one conscious experience.

Some qualia are elemental—the color yellow, the abrupt and overpowering pain of a muscle spam in the lower back, or the feeling of familiarity in *déjà vu*. Others are composites—the smell and feel of my dogs snuggling up against me, the "Aha!" of sudden understanding, or the distinct memory of being utterly transfixed when I first heard the immortal lines: "I've seen things you people wouldn't believe. Attack ships on fire off the shoulder of Orion. I've watched c-beams glitter in the dark near the

Tannhäuser Gate. All those moments will be lost in time, like tears in rain. Time to die." To have an experience means to have qualia, and the qualia of an experience are what specifies that experience and makes it different from other experiences.

I believe that qualia are properties of the natural world. They do not have a divine or supernatural origin. Rather, they are the consequences of unknown laws that I would like to uncover.

Many questions follow from that belief: Are qualia an elementary feature of matter itself, or do they come about only in exceedingly organized systems? Put differently, do elementary particles have qualia, or do only brains have them? Does a single-celled bacterium experience some form of proto-consciousness? What about a worm or a fly? Is there a minimum number of neurons needed for a quale to occur? Or is it the way neurons are wired together that matters most? Can a computer with silicon transistors and copper wire be conscious? Do androids dream of electric sheep, as Philip Dick rhetorically asked? Does my Mac enjoy its intrinsic elegance, whereas my accountant's slab of non-Mac machinery suffers because of its squat gray exterior and clunky software? Is the Internet, with its billion nodes, sentient?

I do not have to start at zero in my quest. We have plenty of facts about consciousness and qualia. Most important, we know that qualia occur in some highly networked biological structures, including the central nervous system of an attentive observer (yours, for example). So the human brain has to be the starting point for any exploration of the physical basis of consciousness.

Not all biological, adaptive, and complex systems qualify, though. Your immune system shows no sign of being conscious. It silently detects and eliminates a wide variety of pathogens day in, day out. Your body's defenses might be engaged, right now, in combating a virus infection without your being aware of it. The immune system will remember this invader and will produce antibodies if the same virus strikes again, giving you lifetime immunity. Yet this memory is not a conscious one.

The same is true of the 100 million neurons that crisscross the inner lining of your gut—the enteric nervous system, sometimes called the second brain. Those neurons go about their business quietly, taking care of nutrient extraction and waste disposal in the gastrointestinal tract. Things you would rather not know about. On occasion, the enteric system acts up—you feel butterflies in your stomach before a crucial job interview or nauseated after a large meal. That information is communicated via the (gastric) vagus nerve to the cerebral cortex, which then generates the nervous or heavy sensation. The second brain in your gut doesn't directly give rise to consciousness.

It is possible that your immune system and enteric nervous system do have a consciousness of their own. Your mind, at home upstairs in the skull, wouldn't know anything about how the gut felt because of the limited communication between the enteric and the central nervous systems. Your body might harbor several autonomous minds, forever isolated, as distant as the dark side of the Moon. Right now, this possibility can't be completely discounted; however, given the limited, stereotyped behaviors of the enteric nervous system, it appears to be subservient to the brain proper, with no capability for independent experience.

Understanding how qualia come about is just the first step toward eliminating the "problem" from the mind–body problem. Next in line is comprehending the particular way that a specific quale feels. Why does red feel the way it does, which is very different from blue? Colors are not abstract, arbitrary symbols: they represent something *meaningful*. If you ask people whether the color orange is situated between red and yellow or between blue and purple, those with normal eyesight will chose the former. There is an inborn organization to color qualia. Indeed, colors can be arranged in a circle, the color wheel. This arrangement is different from that of other sensations, such as the sense of depth or of pitch, which are arranged in a linear sequence. Why? As a group, color percepts share certain commonalities that make them different from other percepts, such as seeing motion or smelling a rose. Why?

I seek a way of answering such questions based on physical principles, on looking at the actual wiring diagram of a brain and deducing from its circuits the sensations that the brain is capable of experiencing. Not just the existence of conscious states, but their detailed character as well. Perhaps you think this feat is beyond science? Don't. Poets, songwriters, and security officials bemoan the impossibility of knowing the mind of anybody else. Although that may be true when looking from the outside, it is not true if I have access to a person's entire brain, with all of its elements. With the correct mathematical framework, I should be able to tell exactly what she or he is experiencing. Like it or not, true mind reading is, at least in principle, possible. More on that in chapters 5 and 9.

What Is the Function of Consciousness?

Another persistent puzzle is why we have any experiences at all. Could we not live, beget, and raise children without consciousness? Such a zombie existence does not contradict any known natural law. From a subjective vantage point, though, it would feel like sleepwalking through life, or like being one of the ghoulish undead in *Night of the Living Dead*. No knowledge. No ego. No being.

What survival value is attached to the inner screen of our mental experiences? What is the function of consciousness, the function of qualia?

The mystery deepens with the realization that much of the ebb and flow of daily life does indeed take place beyond the pale of consciousness. This is patently true for most of the sensory-motor actions that compose our daily routine: tying shoelaces, typing on a computer keyboard, driving a car, returning a tennis serve, running on a rocky trail, dancing a waltz. These actions run on automatic pilot, with little or no conscious introspection. Indeed, the smooth execution of such tasks requires that you *not* concentrate too much on any one component. Whereas consciousness is needed to learn these skills, the point of training is that you don't need to think about them anymore; you trust the wisdom of your body and let it take over. In the words of Nike's ad campaign, you "just do it."

Francis Crick and I postulated the existence of an army of simple-minded *zombie agents* inside each person. These agents are dedicated to stereotypical tasks that can be automated and executed without conscious supervision. You'll read about them in chapter 6. Such nonconscious behavior forced us to examine the benefits of consciousness. Why isn't the brain just a bunch of specialized zombie agents? As they work effortlessly and rapidly, why is consciousness needed at all?

Because life sometimes throws you a curveball! The unexpected happens and you have to think before you act. Your regular route to work is blocked by heavy traffic and you consider alternatives. Francis and I argued that consciousness is important for planning during such episodes: contemplating and choosing a future course of action among many competing ones. Should you wait for the traffic jam to resolve itself, turn left for the long, intersection-free detour or right onto inner city, surface streets with their innumerable traffic lights? Such deliberations require that all relevant facts be summarized and presented to the mind's decision-making stage.

This argument does not imply that all brain activities associated with planning are accessible to consciousness. It may well be that unconscious processes can also plan, but much more slowly than conscious processes or without looking quite as far into the future. Biological systems, unlike artificial ones (think of the electrical wiring in your house) are too redundant, too interwoven to fail completely when one processing module is knocked out. Instead, Francis and I surmised that consciousness permits us to plan more flexibly and further into the future than zombie agents can.

Questions of function are notoriously difficult to answer: Why do we have two eyes rather than eight, as spiders do? What is the function of the vermiform appendix? Rigorously proving why any one particular body feature or behavior arose during evolution is challenging except in general terms, such as eyes are needed to detect prey and predators from a distance.

Some scholars deny that consciousness has any causal role. They accept the reality of consciousness but argue that subjective feelings have no function—they are froth on the ocean of existence, without consequence to the world. The technical term is *epiphenomenal*. The noise the heart makes as it beats is an epiphenomenon: It is useful to the cardiologist diagnosing a patient but of no direct consequence to the body. Thomas Huxley, the naturalist and public defender of Darwin, had this to say in 1884:

The consciousness of brutes would appear to be related to the mechanism of their body simply as a collateral product of its working, and to be completely without any power of modifying that working as the steam-whistle which accompanies the work of a locomotive engine is without influence upon its machinery.

I find this line of argument implausible (yet cannot refute it for now). Consciousness is filled with meaningful percepts and memories of sometimes unbearable intensity. Why would evolution have favored a tight and consistent link between neural activity and consciousness, a link that persists throughout a lifetime, if the feeling part of this partnership didn't have any consequences for the survival of the organism? Brains are the product of a selection process that has operated over hundreds of million of iterations; if qualia didn't have any function, they would not have survived this ruthless vetting process.

While speculations about the function of consciousness engage the imagination of philosophers, psychologists, and engineers, the empirical inquiry into its material basis progresses at a dizzying pace. Science is best at answering mechanistic "How?" rather than final "Why?" questions. Obsessing about the utility of consciousness is less fruitful as an avenue of research than studying which bits of the brain are important for it.

The Difficulty of Defining Consciousness

After lecturing about consciousness in animals, I was approached by a lady who emphatically exclaimed, "You will never be able to convince me that a monkey is conscious!" "And you can never convince me that *you* are conscious," I retorted. She was taken aback by my response, but then the light of understanding dawned on her face: The inability to feel what a monkey, a dog, or a bird feels applies to people as well, though to a lesser degree. Consider a spy working undercover, a cheating lover, or a professional actor. They fake feelings of trust, patriotism, love, or friendship. You can never be absolutely confident about the feelings of anybody! You can watch their eyes, parse what they say, but in the end you cannot truly know their mind by observing them.

Defining what, exactly, constitutes jazz is notoriously difficult. Hence the saying, "Man, if you gotta ask, you're never gonna know." The same holds true for consciousness. It isn't possible to explain conscious feelings without invoking other feelings, qualia, or states of sentience. This difficulty is reflected in the many circular definitions of consciousness. The *Oxford English Dictionary,* for instance, defines consciousness as "the state of being conscious, regarded as the normal condition of healthy waking life." Consider a related but simpler problem: explaining to a person blind from birth what color is. Defining the percept of red is impossible without referring to other red objects. And the color of an object is different from its shape, weight, or smell. Its color is similar only to, well, other colors.

Definitions can be dangerous if they are attempted before having reached a thorough understanding: Premature rigor can act as a Procrustean bed, constraining progress. If I seem evasive, try defining a *gene*. Is it a stable unit of hereditary transmission? Does it have to code for a single enzyme? What about structural and regulatory genes? Does a gene correspond to one continuous segment of nucleic acid? Then what about introns, splicing, and posttranslational editing? Or try something simpler, such as defining a *planet*. I grew up learning about nine planets. A couple of years ago, my Caltech colleague Michael Brown led a team that discovered Eris, a planet in the distal reaches of the solar system. Eris has its own moon and is more massive than Pluto. Because it is also one of a bevy of so-called trans-Neptunian objects, astronomers started scratching their heads. Did this mean there was going to be an endless number of new planets? To avoid this, they redefined what a planet is and demoted Pluto to the status of a dwarf, or minor, planet. If astronomers sight planets orbiting a binary star system, or solitary planets that wander through space without a companion star to light their sky, the definition of a planet will have to adapt once more.

A habitual misperception is that science first rigorously defines the phenomena it studies, then uncovers the principles that govern them. Historically, progress in science is made without precise, axiomatic formulations. Scientists work with malleable, *ad hoc* definitions that they adapt as better knowledge becomes available. Such working definitions guide discussion and experimentation and permit different research communities to interact, enabling progress.

In this spirit, let me offer four different definitions of consciousness. Like the Buddhist fable of the blind men, each describing different aspects of the same elephant, each captures an important facet of consciousness, with none of them painting a complete picture.

A *commonsense definition* equates consciousness with our inner, mental life. Consciousness begins when we wake up in the morning and continues throughout the day until we fall into a dreamless sleep. Consciousness is present when we dream but is exiled during deep sleep, anesthesia, and coma. And it is permanently gone in death. *Ecclesiastes* had it right: "For the living know that they shall die but the dead know not anything."

A *behavioral definition* of consciousness is a checklist of actions or behaviors that would certify as conscious any organism that could do one or more of them. Emergency room personnel quickly assess the severity of a head injury using the *Glasgow Coma Score*. It assigns a number to the patient's ability to control their eyes, limbs, and voice. A combined score of 3 corresponds to coma, and a 15 marks the patient as fully conscious. Intermediate values correspond to partial impairment: for instance, "is confused and disoriented in response to questioning but withdraws from painful stimuli." Although this measure of consciousness is perfectly reasonable when dealing with children or adults, the challenge is to come up with behavioral criteria appropriate to organisms that can't readily be asked to do something, such as babies, dogs, mice, or even flies.

A further difficulty lies in distinguishing stereotypical, automatic reflexes from more complicated actions that demand consciousness (I'll return to such zombie behaviors in chapter 6). In practice, if a subject repeatedly behaves in a purposeful, nonroutine manner that involves the brief retention of information, she, he, or it is assumed to be conscious. If I make a face and stick out my tongue at an infant and the infant mimics me a bit later, then it is fair to assume that the baby has at least some rudimentary awareness of its surroundings. Likewise, if a patient is asked to move her eyes to the left, to the right, and then up and down, and she complies, then she is conscious. Again, such tasks have to be adapted to the hands, paws, pads, claws, flippers, or snouts of the species being tested.

A *neuronal definition* of consciousness specifies the minimal physiologic mechanisms required for any one conscious sensation. Clinicians know, for example, that if the brain stem is impaired, consciousness is dramatically reduced and may be absent altogether, leading to a vegetative state. Another condition necessary for any one specific conscious sensation is an active and functioning *cortico-thalamic complex*. This complex includes, first and foremost, the neocortex and the closely allied thalamus underneath it. The neocortex is the newest part of the cerebral cortex, the folded layers of neurons that constitute the proverbial gray matter. It occupies most of the forebrain and is a unique hallmark of mammals. The thalamus is a quail egg–sized structure in the middle of the brain that regulates all input into the neocortex and receives massive feedback from it. Pretty much every region of the cerebral cortex receives input from a specific region of the thalamus and sends information back to it. Other structures that are lumped in with the cortico-thalamic duo—hence the moniker "complex"—are the hippocampus, amygdala, basal ganglia, and claustrum. Most progress in the science of consciousness has occurred in the neuronal domain. I'll go into more details in chapters 4 and 5.

If these three definitions don't nail the problem, ask a *philosopher*. She'll give you a fourth definition, "consciousness is what it is like to feel something." What it feels like to have a specific experience can only be known by the organism having the experience. This what-it-feels-like-from-within perspective expresses the principal, irreducible trait of phenomenal awareness—to experience something, anything.

None of these definitions is foundational. None describes in unequivocal terms what it takes for any system to be conscious. But, for practical purposes, the behavioral and neuronal definitions are the most useful.

On the Consciousness of Animals

To progress toward an answer to these difficult questions without getting bogged down in diversionary skirmishes and heated debates, I have to make some assumptions without being able at this stage to fully justify them. These working hypotheses may well need to be revised, or even rejected, later on.

As a starting point, I assume that the physical basis of consciousness is closely linked to specific interactions among neurons and their elements. Although consciousness is fully compatible with the laws of physics, it is not feasible to predict or understand consciousness from these laws alone.

I furthermore assume that many animals, mammals especially, possess some of the features of consciousness: They see, hear, smell, and otherwise experience the world. Of course, each species has its own unique sensorium, matched to its ecological niche. But they all experience something. To believe otherwise is presumptuous and flies in the face of all evidence for the structural and behavioral continuity between animals and humans. Biologists emphasize this continuity by distinguishing between *non-human* and *human animals*. We are all nature's children. I am confident of this for three reasons.

First, the behavior of many mammals is kindred, though not identical. Take my dog when she yelps, whines, gnaws at her paw, limps, and then comes to me, seeking aid: I infer that she's in pain because under similar conditions I behave in similar ways (*sans* gnawing). Physiologic measures of pain confirm this inference—dogs, just like people, have an elevated heart rate and blood pressure and release stress hormones into their bloodstream. And it is not only physical pain that we share with animals but suffering as well. Suffering occurs when animals are systematically abused or when an older pet is separated from its litter mate or its human companion. I'm not saying that dog-pain is exactly like people-pain, but that dogs—and other animals as well—not only react to noxious stimuli but also consciously experience pain.

Second, the structure of the nervous system is comparable across mammals: It takes an expert neuroanatomist to distinguish between a pea-sized chunk of cerebral cortex taken from a mouse, a monkey, and a person. Our brain is big, but other creatures—elephants, dolphins, and whales—have bigger ones. Neither at the genomic nor at the synaptic, cellular, or connectional levels are there qualitative differences between mice, monkeys, and people. The receptors and pathways that mediate pain are analogous across species. After an extensive trail run, I give my dog the same pain medication that I take (in a dose appropriate to her body size) and her limp goes away. Hence, I suppose that dog-pain feels bad, just as people-pain does.

Despite these similarities, there are a myriad of quantitative differences at the hardware level. They add up to the fact that as a species, *Homo sapiens* can build a globe-spanning Internet, plan for nuclear war, and wait for Godot, feats that other animals are incapable of. But no qualitative brain differences have been uncovered.

Third, all extant mammals are closely related to each other. Placental mammals evolved into the diverse forms seen today after the extinction of the dinosaurs by the fiery impact of an asteroid in the Yucatán some 65 million years ago. Great apes, chimpanzees, and gorillas shared a common ancestor with humans as recently as six million years ago. *Homo sapiens* is part of an evolutionary continuum, not a unique organism that dropped, fully sentient, from the sky.

It is possible that consciousness is common to *all* multicellular animals. Ravens, crows, magpies, parrots, and other birds; tuna, cichlid, and other fish; squid; and bees are all capable of sophisticated behavior. It is likely that they too possess some inkling of awareness, that they suffer pain and enjoy pleasure. What differs among species and even among members of the same species is how differentiated, how braided and complex these conscious states can be. What they are conscious of—the content of their awareness—is closely related to their senses and their ecological niches. To each its own.

The repertoire of conscious states must somehow diminish with the diminishing complexity of an organism's nervous system. Whether the two most popular species in biology labs—the roundworm *Caenorhabditis elegans*, with its 302 nerve cells, and the fruit fly *Drosophila melanogaster*, with its 100,000 neurons—have any phenomenal states is difficult to ascertain at the moment. Without a sound understanding of the neuronal architecture necessary to support consciousness, we cannot know whether there is a Rubicon in the animal kingdom that separates sentient creatures from those that do not feel anything.

On the Consciousness of Self

Ask people what they believe to be the defining feature of consciousness and most will point to self-awareness. To be conscious of yourself, to worry about your child's illness, to wonder why you feel despondent or why she provoked your jealousy is taken to be the pinnacle of sentience.

Young children have only very limited insight into their actions. If they are younger than eighteen months, they do not recognize themselves in a mirror. Behavioral psychologists use such a *mirror test* as gold standard for self-recognition. The infant is clandestinely marked with a spot or color patch on the forehead or face. Exposed to a familiar mirror, the baby will play with its mirror image but won't scratch or try to remove the mark from its face, unlike teenagers who can endlessly hog the bathroom, fussing about their appearance. A number of species besides humans pass the mirror test (with appropriate modifications): great apes, dolphins, elephants, and magpies. Monkeys bare their teeth or otherwise interact with their reflections, but they do not realize that "the image there corresponds to my body here." That is not to say that they have no sense of self, but at least not a visual representation of their body that they compare against the external view in the mirror.

Some scholars conclude from this lack of self-awareness that the vast majority of the animal kingdom is unconscious. By this measure, only humans, and not even their young ones, are truly conscious.

One among many observations makes this conclusion implausible. When you are truly engaged with the world, you are only dimly aware of yourself. I feel this most acutely when I climb mountains, cliffs, and desert towers. On the high crag, life is at its most intense. On good days, I experience what the psychologist Mihaly Csikszentmihalyi calls *flow*. It is a powerful state in which I am exquisitely conscious of my surroundings, the texture of the granite beneath my fingers, the wind blowing in my hair, the Sun's rays striking my back, and, always, always, the distance to the last hold below me. Flow goes hand-in-hand with smooth and fluid movements, a seamless integration of sensing and acting. All attention is on the task at hand: The passage of time slows down, and the sense of self disappears. That inner voice, my personal critic who is always ready to remind me of my failings, is mute. Flow is a rapturous state related to the mind-set of a Buddhist lost in deep meditation.

The writer and climber Jon Krakauer puts it well in *Eiger Dreams*:

By and by, your attention becomes so intensely focused that you no longer notice the raw knuckles, the cramping thighs, the strain of maintaining nonstop concentration. A trance-like state settles over your efforts, the climb becomes a clear-eyed dream. The accrued guilt and clutter of day-to-day existence, all of it is temporarily forgotten, crowded from your thoughts by an overpowering clarity of purpose, and by the seriousness of the task at hand.

This loss of self-awareness not only occurs in alpinism but also when making love, engaging in a heated debate, swing dancing, or racing a motorcycle. In these situations, you are in the here and now. You are in the world and of the world, with little awareness of yourself.

Rafael Malach at the Weizmann Institute in Israel paid volunteers to watch *The Good, The Bad and The Ugly* while lying on their back inside the narrow confines of a brain-imaging machine. Though not your average cinematic experience, the viewers nonetheless found this classic Spaghetti Western engaging. When analyzing their brain patterns, Rafi found that regions of the cerebral cortex associated with introspection, high-level cognition, planning, and evaluation were relatively deactivated, but the regions involved in sensory, emotional, and memory processes were furiously busy. What's more, the rise and fall of the blood flow in the cortex, tracked by the brain scanner, was similar across volunteers. Both observations reflect the mastery with which the gifted Italian director Sergio Leone controls his audience, making them see, feel, and recall what he wants them to see, feel, and recall. That is one reason we like to watch movies—they distract us from our overly active self-consciousness, from our daily barrage of worries, anxieties, fears, and doubts. For a few hours, we escape the tyranny of our skull-sized kingdoms. We are highly conscious of the events in the story, but only slightly aware of our own internal state. And that is sometimes a great blessing.

As it happens, people with widespread degeneration of the front of the cerebral cortex have substantial cognitive, executive, emotional, and planning deficits, coupled with a lack of insight into their abysmal condition. Yet their perceptual abilities are usually preserved. They see, hear, and smell, and are aware of their percepts.

Self-consciousness is part and parcel of consciousness. It is a special form of awareness that is not concerned with the external world, but is directed at internal states, reflections about them, and reflections upon such reflections. This recursiveness makes it a peculiarly powerful mode of thinking.

The computer scientist Doug Hofstadter speculates that at the heart of self-consciousness is a paradoxical and self-referential *strange loop*, akin to the Escher print of a pair of hands drawing each other. If so—and I'm skeptical, because the mind has difficulty self-referencing or recursing much beyond "I am thinking about myself thinking"—this strange loop is incidental to consciousness per se. Rather, self-consciousness is an evolutionary adaptation of older forms of body and pain consciousness.

Another singular human trait is speech. True language enables *Homo sapiens* to represent, manipulate, and disseminate arbitrary symbols and concepts. Language leads to cathedrals, the Slow Food movement, the theory of general relativity, and *The Master and Margarita*. Such things are beyond the capabilities of our animal friends. The primacy of language for most aspects of civilized life has given rise to a belief among philosophers, linguists, and others that consciousness is impossible without language and, therefore, that only humans can feel and introspect.

I defy this view. There is little reason to deny consciousness to animals because they are mute or to premature infants because their brain is not fully developed. There is even less reason to deny it to people with severe aphasia, who upon recovery can clearly describe their experiences while they were incapable of speaking. The perennial habit of introspection has led many intellectuals to devalue the unreflective, nonverbal character of much of life and to elevate language to the role of kingmaker. Language is, after all, their major tool.

What about emotions? Does an organism have to feel angry, scared, disgusted, surprised, sad, or excited to be conscious? Although there is no doubt that such strong feelings are essential for our survival, there is no convincing evidence that they are essential for consciousness. Whether or not you are angry or happy, you will continue to see the candle burning in front of you and will feel the pain when you put your finger into its flame.

Some people have flat affect due to severe depression or damage to the frontal lobes of their brain; their actions are impaired and their judgment unsound. The head-injured war veteran can recall with nonchalance the blast of the mine that exploded under his Humvee and took his legs. He seems completely aloof, detached from, and disinterested in his condition, yet there is no doubt that he is experiencing something, if only extreme discomfort from his injuries. Emotions are indispensable to a well-balanced and successful life but are not essential for consciousness.

We're done with the brush-clearing that prepares the true battleground on which the mind–body problem will be resolved. The patient I just mentioned reminds us of the centrality of the brain. Neuroscience textbooks describe this organ in mind-numbing detail yet leave out what it feels like to be the owner of one. Let me make up for this remarkable omission by linking the interior view of the experiencing subject with the exterior perspective of the brain scientist.

Chapter 4: In which you hear tales of scientist-magicians that make you look but not see, how they track the footprints of consciousness by peering into your skull, why you don't see with your eyes, and why attention and consciousness are not the same

"Data! Data! Data!" he cried impatiently. "I can't make bricks without clay."
—Sherlock Holmes in *The Adventure of the Copper Beeches*, Sir Arthur Conan Doyle (1892)

People willingly concede that when it comes to nuclear physics or kidney dialysis, specialized knowledge is essential. But let the conversation turn to consciousness, and everybody chimes in, on the assumption that they are all entitled to their own pet theory in the absence of pertinent facts. Nothing could be further from the truth.

An immense stockpile of psychological, neurobiological, and medical knowledge about the brain and the mind has accumulated. The travails of more than fifty thousand brain and cognitive scientists worldwide add thousands of new studies each year to this vast collection.

But don't worry. I won't pontificate on even a fraction of this material. I'll focus, instead, on a few vignettes from the laboratory that characterize the modern quest for the roots of consciousness.

Finding Consciousness in the Brain

In the closing decade of the past millennium, a few intrepid scientific explorers, tired of the endless, eristic metaphysical arguments—does consciousness really exist? is it independent of the laws of physics? how does the intentionality of mental states come about? which of the many "-isms" invented by philosophers in their carefree ways best describes the relationship between the mind and the body?— looked for the footprints of consciousness in the brain. This can-do attitude resulted in a great conceptual advance—consciousness began to be conceived of as the product of particular brain mechanisms.

In the early 1990s, Francis Crick and I focused on what we called the *neural* (or *neuronal*) *correlates of consciousness* (NCC). We defined them as the minimal neural mechanisms jointly sufficient for any one specific conscious percept. (Our definition does not speak to the possibility of creating consciousness in machines or in software, a deliberate omission to which I shall return later on.)

Imagine that you are looking at a red cube, mysteriously left in the desert sand, with a butterfly fluttering above it. Your mind apprehends the cube in a flash. It performs this feat because the brain activates specialized cortical neurons that represent color and combines them with neurons that encode the percept of depth, as well as neurons that encode the orientation of the various lines that make up the cube. The minimal set of such neurons that causes the conscious percept is the neural correlate of consciousness for perceiving this alien object.

It is important to stress the "minimal." Without that qualifier, all of the brain could be considered a correlate: After all, the brain does generate consciousness, day in and day out. But Francis and I were after bigger game—the specific synapses, neurons, and circuits that generate, that cause, consciousness. Being fastidious scientists, we used the more cautious "correlates" in lieu of the more definite "causes" of consciousness.

Some sectors of the brain have a more intimate, a more privileged relationship to the content of consciousness than others. The brain is not like a hologram, in which everything contributes equally to the image. Some regions add little, if anything, and can be damaged without loss of phenomenal experience, whereas others are crucial to consciousness.

Do you really need your spinal cord to see consciously? Hemiplegics and quadriplegics lose feeling in and control of their bodies below their spinal cord injury, yet they are clearly aware of the world and can lead lives as rich and meaningful as any. Think of Christopher Reeve, the actor who portrayed Superman. Paralyzed from the neck down after a horse-riding accident, he started a medical foundation and became an eloquent spokesman for stem cell research and other rehabilitation techniques that will enable quadriplegics like himself to regain mobility some day.

What about the cerebellum, the little brain at the back of the head, underneath the cerebral cortex? The moniker "little" is ironic, because the cerebellum has 69 billion nerve cells, more than four times the number in the celebrated cerebral cortex, which hogs all of the limelight. If a stroke or a tumor strikes your cerebellum, your equilibrium and coordination are impaired. Your gait is clumsy, your stance wide, you shuffle about, your eye movements are erratic, and your speech is slurred. The smooth and refined movements that you took for granted are jerky now and have to be laboriously and consciously willed. Playing the piano or a tennis match becomes a distant memory. Yet your perceptions and memories do not change much, if at all. Sound, sight, touch and smell remain unaffected.

Contrast damage to the spinal cord or cerebellum with the effects of a lesion — a circumscribed piece of tissue destroyed by some process — in the cerebral cortex or the hippocampus. Depending on where the destruction occurs, you may be unable to recall anything, even the names of your children; colors may be leached from the world; or you may be unable to recognize familiar faces. From such clinical observations, described more fully in the next chapter, scientists have concluded that bioelectrical activity in discrete regions of the cerebral cortex and its satellites is essential for the content of conscious experience.

The question of which brain regions are essential for consciousness is hotly debated. The neurologist Antonio Damasio at the University of Southern California argues that parts of the parietal lobe in the back of the cerebral cortex are essential. Others point to the "anterior insula," to the "superior temporal polysensory" cortex, or to other regions of the brain. In the fullness of time, a list of all regions that are both *necessary* and *sufficient* for consciousness is needed, but for now that remains a distant goal.

Identifying a specific region of the brain as critical is just the beginning. It's like saying that the murder suspect lives somewhere in the Northeast. It's not good enough. I want to home in on the specific circuits, cell types, and synapses in those brain regions that are key to mediating a specific experience. Francis and I proposed that a critical component of any neural correlate of consciousness is the long-distance, reciprocal connections between the higher-order sensory regions, located in the back of the cerebral cortex, and the planning and decision-making regions of the prefrontal cortex, located in the front. I'll return to this idea later.

Every phenomenal, subjective state is caused by a particular physical mechanism in the brain. There is a circuit for seeing your grandmother in a picture or in life, another one for hearing the sound of the wind whispering through pine trees on a mountaintop, and a third one for the vicarious rush when rapidly weaving on a bicycle through city traffic.

What are the commonalities among the neural correlates for these distinct qualia? Do they all involve the same circuit motifs? Or, do they all involve neurons in area X? Are all three sensations mediated by active pyramidal neurons in the prefrontal cortex that snake their output wires back to the relevant sensory regions? Do the neurons that mediate phenomenal content fire in a rhythmic and highly coordinated manner? These are some of the ideas that Francis and I have entertained.

Perturbing the neural correlate of any specific conscious experience will change the percept. Destroying or otherwise silencing the relevant neurons will make the percept disappear, although the brain, especially a young one, can compensate for limited damage within weeks.

Artificially inducing a neural correlate of consciousness will trigger the associated percept. This is done routinely in neurosurgery: The surgeon places an electrode onto the surface of the brain and sends an electric current through it. Depending on the location and intensity, this external stimulus can trigger a poignant memory, a song last heard years before, the feeling of wanting to move a limb or the sensation of movement.

The limit of the possible is explored in the "edge of the construct" movie *The Matrix*. With the help of an electro-organic link interfaced via jacks to the back of the head and spinal cord, the Machines create an entirely fictitious and wholesome world in Neo's mind by stimulating the appropriate neural correlates of consciousness. Only by unplugging from this stimulator does Neo "wake up to" the fact that he is vegetating his life away in a gigantic stack of cages, cared for by insect-like robots that breed humans to obtain energy from their bodies.

You, too, hallucinate every night in the privacy of your head. During sleep, you have vivid, sometimes emotionally wrenching, phenomenal experiences, even if you don't recall most of them. Your eyes are closed, yet the dreaming brain constructs its own reality. Except for rare "lucid" dreams, you can't tell the difference between dreaming and waking consciousness. Dreams are real while they last. Can you say more of life?

Ironically, while in your mind you act out your dreams, your sleeping body is largely paralyzed: Your brain restricts bodily movement to protect itself from the sometimes violent movements that you dream about. This paralysis implies that behavior is not really necessary for consciousness. The adult brain, even if cut off from most input and output, is all that is needed to generate that magical stuff, experience. The old philosopher's standby—the brain in a vat, which saw a modern renaissance in *The Matrix*—suffices.

Discovering and characterizing the neural correlates of consciousness by homing in on the relevant neuronal circuits is the theme of much of contemporary research, especially research on vision.

Hiding Objects in Plain Sight

A couple of years ago, I spent a few days in New York City with an accomplished thief, Apollo Robbins. Based in Las Vegas, he is a professional magician and excels at all the usual feats of stage prestidigitation and conjuring. Yet he was at his most compelling when seated next to me in a café—bereft of fog, spotlights, bikini-clad assistant, and music, all of which serve to distract the viewer. He pulled coins out of thin air, threw paper balls at me that abruptly disappeared, and stole my watch— all while I was watching his every movement. And I call myself an expert in visual perception! One of Apollo's most impressive acts was when he took a card from my son and stuck it to his own forehead. The card was there in plain sight, yet my son was mystified as to where it had gone because his attention was riveted on the magician's hand.

What I learned during my time with Apollo and others like him is that thieves and magicians are masters at manipulating the attention and the expectations of their audience. If Apollo can misdirect your gaze or your attention to his left hand, then you will be blind to the actions of his right hand, even though you're looking at it.

The spatial focus of your attention is referred to as the *spotlight of attention*. Whatever object or event is illuminated by this inner lantern is treated preferentially and can be detected faster and with fewer errors. Of course, where there is light, there is also shade—objects or events that are not attended are often only barely perceived.

Let's transition from a busy Manhattan café to the inside of the claustrophobic, coffin-like confines of a loud and pounding magnetic resonance imaging (MRI) scanner, a massive machine that weighs several tons. You are lying inside a cramped cylinder, desperately trying to keep still, to not even bat an eyelid, as any movement causes the signals to wash out. Through a mirror, you stare at a computer monitor displaying the ace of hearts from a deck of cards while the machine monitors blood flow in your brain. Neuroscientists are not good at sleight of hand, so they manipulate what you see by projecting a precisely timed second image into your eyes. If done well, this misdirection works as well as the magician's—you won't see the ace of hearts. The second image *masks* the first one, rendering the ace invisible. You look, but don't see, inverting Yogi Berra's famous witticism "You can observe a lot by watching."

This technique was perfected by my then-graduate student Naotsugu Tsuchiya who called it *continuous flash suppression*. It works by projecting the image of the playing card into one of your eyes while continuously flashing a multitude of brightly colored, overlapping rectangles—like those of the Dutch painter Piet Mondrian—into the other eye. If you wink with this eye, the ace of hearts becomes visible, but if you keep both eyes open, the ace remains hidden for minutes on end, camouflaged by the ever-changing display of colored rectangles that distracts you.

Such powerful masking techniques are one reason why the study of vision is blooming. With them, it is comparatively easy to make people look but not see, to manipulate what they perceive—far easier than to trick the other senses. The sense of smell and the sense of who-you-are are more robust, less prone to manipulation. I will not be able to make you confuse the smell of a rose with that of blue Stilton cheese or make you believe that you are the Queen of England one moment and Madonna the next.

The best experiment is one in which everything but the variable of interest is kept fixed. That way, the difference this one thing makes to the action of the entire system can be isolated. Neuroscientists use a magnetic scanner to compare your brain's activity when you see the ace of hearts with brain activity when you are looking at it without seeing the ace because it is masked. By considering the difference, they can isolate the activity that is unique to the subjective experience of seeing the ace, tracking consciousness to its lair by following its footprints.

The image of the ace of hearts stimulates neurons in your eye, called retinal ganglion cells. Their axons make up the optic nerve, which sends visual information to the brain proper. These cells respond to the ace of hearts with a burst of action potentials, the brief, all-or-none pulses described in chapter 2. The output of the eye does not depend on whether the owner of the eyeball is conscious. As long as the eyelids are open, the optic nerve faithfully signals what is out there and passes this on to downstream structures in the cerebral cortex. This activity ultimately triggers the formation of a stable coalition of active cortical neurons that conveys the conscious percept of a red ace. How this comes about is the topic of chapter 8.

No coalition can form, however, in the face of competition with the much more vigorous waves of spikes that roll in unceasingly on the optic nerve of the other eye, which is being stimulated by constantly changing colored rectangles. These spikes trigger their own neural coalition, and as a result, you see flashing colored surfaces, while the red ace remains invisible.

To understand the next experiments, let me briefly explain the principle underlying nuclear magnetic resonance imaging (the word "nuclear" has been dropped because of its negative connotation). The MRI scanner generates a powerful magnetic field, about 100,000 times stronger than Earth's magnetic field. The nuclei of certain elements, including hydrogen, behave like miniature bar magnets. When you enter a scanner's strong magnetic field, the hydrogen nuclei in your body line up with this field. More than half of your body weight is water, which is made up of two hydrogen atoms and one oxygen atom. The MRI scanner sends a brief pulse of radio waves into your skull, knocking the nuclei out of alignment. As the nuclei relax back into their original state, they give off faint radio signals that are detected and processed into a digital image. Such a picture reveals the structure of soft tissue. For example, it demarks the boundary between the brain's gray and white matter. Magnetic resonance imaging is much more sensitive than X-rays. It has revolutionized medicine by allowing tissue damage, from a tumor to trauma, to be located and diagnosed with negligible risk to the patient.

Whereas MRI renders the internal structure of organs visible, functional MRI (fMRI) relies on a fortuitous property of the blood supply to reveal regional brain activity. Active synapses and neurons consume power and therefore need more oxygen, which is delivered by the hemoglobin molecules inside the circulating red blood cells. When these molecules give off their oxygen to the surrounding tissue, they not only change color—from arterial red to venous blue—but also turn slightly magnetic. Activity in neural tissue causes an increase in the volume and flow of fresh blood. This change in the blood supply, called the hemodynamic signal, is tracked by sending radio waves into the skull and carefully listening to their return echoes. Keep in mind that fMRI does not directly measure synaptic and neuronal activity, which occurs over the course of milliseconds. Instead, it uses a relatively sluggish proxy—changes in the blood supply—that rises and falls in seconds. The spatial resolution of fMRI is limited to a volume element (voxel) the size of a pea, encompassing about one million nerve cells.

So, how does the brain respond to things the mind doesn't see?

Remarkably, invisible pictures can leave traces in cortex. Wisps of unconscious processing can be picked up in the primary visual cortex (V1). The primary visual cortex is the terminus for information sent from the eyes. Located just above the bump at the back of the head, it is the first neocortical region evaluating pictorial information. Other sectors of the cerebral cortex respond to suppressed pictures as well—notably the cascade of higher-order visual regions (V2, V3, and so on) that extends beyond the primary visual cortex, and the amygdala, the almond-sized structure that deals with emotional stimuli such as fearful or angry faces.

As a rule, the farther removed a region of the visual brain is from the retina, the more strongly felt is consciousness' influence. As expectations, biases, and memory come to play a larger role in higher regions of the brain, the impact of the external world weakens correspondingly. The subjective mind manifests itself most strongly in the upper echelon of the cortex. That is its habitat.

This implies that not just any cortical activity is sufficient for conscious sensation. Even though a million neurons in the primary visual cortex are busily firing away, their exuberant spiking may not give rise to an experience, if higher-up neurons are not reflecting this activity. More is needed. Maybe this activity has to exceed some threshold? Maybe a specific set of special neurons in the back of the cortex have to establish a reciprocal dialogue with neurons in the frontal lobes? Maybe these neurons have to be jointly active, firing action potentials in unison like synchronized swimmers? I suspect all three conditions must be met for information to be perceived consciously. Such conclusions have stoked the fires of a lively debate concerning the extent to which neurons early on in the visual processing hierarchy are responsible for generating conscious percepts.

In a noteworthy experiment by Sheng He and his group at the University of Minnesota, volunteers looked at a photo of a naked man or woman in one eye while colored rectangles were continuously flashed into the other eye, rendering the nudes invisible. Nonetheless, sensitive tests indicated that invisible pictures of naked women attracted the attention of straight men, whereas images of naked men repulsed them. Yet this occurred *sub rosa*, under the radar of consciousness. The volunteers didn't see the nude yet still paid attention. Conversely, the attention of straight women—and of gay men—was attracted by invisible pictures of naked men. Functionally, this makes sense, as your brain needs to know about potential mates. It also affirms a widespread cliché about the unconscious nature of desire.

Not all Neurons Participate in Consciousness

In 1995, Francis and I published a manuscript in the international journal *Nature* with the title, "Are we aware of neural activity in primary visual cortex?" (Having your article appear in *Nature* is like having the premier gallery in New York or Paris display your art; it's a big deal.) We answered our rhetorical question in the negative, arguing that the neural correlates of visual percepts are not to be found there. We based this hypothesis on the neuroanatomy of the macaque monkey.

Both monkeys and humans belong to the primate order. Our last common ancestor lived between 24 million and 28 million years ago. The visual system of monkeys is similar to ours; they adapt well to captivity; they can be trained easily; and they are not an endangered species. Thus, macaque monkeys are the species of choice for scientists interested in higher-order perception and cognition. For obvious ethical reasons, the human brain is not readily accessible to invasive probing. As a consequence, we know much more about the wiring in the monkey than in the human brain.

Pyramidal neurons are the workhorses of the cerebral cortex. They account for about four of every five cortical neurons and are the only ones that convey information from one region to other sectors within or outside of the cortex, such as the thalamus, the basal ganglia, or the spinal cord. Pyramidal cells in the primary visual cortex send their information to other regions, such as V2 and V3, but none of them reaches all the way to the front of the brain. It is there, in the prefrontal cortex, and especially in its dorsolateral division, that the higher intellectual functions—problem solving, reasoning, and decision making—are located. Damage to this region will leave a person's sensory modalities and memory intact, but her capacity to make rational decisions will be compromised: She will engage—and perseverate—in highly inappropriate behavior.

In the past chapter, I mentioned Francis and my hunch that the function of consciousness is planning. Patients bereft of part or all of their prefrontal cortex have difficulty planning for the near or the distant future. We took this to imply that the neural correlates of consciousness must include neurons in the prefrontal cortex. And because neurons in the primary visual cortex do not send their axons that far forward, we concluded that neurons sited in the primary visual cortex were not part and parcel of the neurons that underpin visual consciousness.

Our proposal is counterintuitive because, under many conditions, the bioelectrical activity of neurons in the primary visual cortex does reflect what a person sees. Take right now, as you're reading these words: The strokes making up the letters excite retinal neurons that convey their content to the primary visual cortex. From there, the information is relayed to the cortical *visual word form* area. Your ability to see these letters seems to disprove our hypothesis—the activity of neurons in the primary visual cortex can correlate with conscious perception. The same is also true for retinal neurons—their activity can mirror what you see.

Although the responses of cells in the retina and primary visual cortex do share certain attributes with visual consciousness some of the time, during other conditions their responses can differ greatly. Let me give you three examples of why you don't see with your eyes—something that painters have known for centuries.

First, consider the incessant movements of your eyes as you skim these lines. You make several rapid eye movements, called saccades, each second. Despite the nearly continual motion of your image sensor, the page looks steady. This should surprise you. As you saccade to the right, the world should shift leftward—but it doesn't! Consider what would happen if you were to move a video camera with the same cadence across the book: watching the resultant movie would make you nauseated. To avoid this, television cameras pan slowly across a scene; their motion is completely different from the way your eyes dart about as you take in salient aspects of an image. If retinal neurons were the ones communicating the percept of a stationary world, they would have to exclusively signal motion in the outside world, but not respond to motion of the eye in which the cells are themselves located. However, retinal nerve cells, like those in the primary visual cortex, are incapable of distinguishing between object motion and eye motion. They react to both. Unlike smart phones, they do not have an accelerometer or a GPS sensor attached to them that distinguishes external, image motion from internal, ocular motion. It is neurons in the higher reaches of the visual cortex that produce your perception that the world is stationary.

Second, consider the "hole" in your retina at the *blind spot*: This is the place where the optic nerve leaves the retina; the axons making up the nerve fiber displace the photoreceptors, so none of the incoming photons are captured at the blind spot. If you were conscious of the informational content of your retinal cells, you wouldn't make out anything there, just as you don't see the boundaries of your field of view on your extreme left and right sides. You would be annoyed if the camera on your cell phone contained a few bad pixels that were always off: The black spot in the photos would drive you nuts. But you don't perceive the hole in your eye because cortical cells actively fill in information from the margins, compensating for the absence of information.

As a third piece of evidence, recall that the world of your dreams is colored, animated, and fully rendered. Because you sleep in the dark, with your eyes shut, the nerve cells in your eyes don't signal anything about the outside world. It is the cortico-thalamic complex that provides the phenomenal content of dreams.

These are just three of numerous dissociations between the state of retinal neurons and what you consciously see. The myriad action potentials streaming up your optic nerve carry data that are heavily edited before they become part of the neural correlates of consciousness. And sometimes retinal information is dispensed with altogether, such as when you close your eyes and conjure up Winnie the Pooh, your inseparable childhood companion, or when you dream about him.

Similar objections can be made to the proposition that cells in the primary visual cortex directly contribute to visual perception. Take recordings of action potentials made with fine electrodes in the macaque monkey's brain. Once the skull has been breached via a small burr hole drilled while the animal is under anesthesia, microelectrodes are slowly inserted into the gray matter and are connected to amplifiers. The nervous tissue itself has no pain receptors, so once the electrodes are in place, they don't cause discomfort. (Think of the half million patients who have electrodes implanted into their brain to ameliorate a variety of illnesses, such as the shaking of Parkinson's disease.) The electrodes pick up the faint electrical pulses emitted by nerve cells. These signals can be played back through a loudspeaker. You can clearly hear the staccato sounds of spikes above the background hiss of the nervous system humming to itself. Such recordings confirm that neurons in the primary visual cortex respond to both motion of the monkey's eyes and to movement of the image. When the monkey's eyes move, these neurons all sound off in unison, signaling change. Yet as I just discussed, your world looks steady as you move your eyes about.

Hemodynamic activity in the primary visual cortex often reflects what a person sees, but it can sometimes also be strikingly disconnected. John-Dylan Haynes and Geraint Rees at University College London briefly flashed stripes tilted either to the left or to the right into the eyes of volunteers; these stripes were then masked so that the observers couldn't tell in which direction they were oriented. All they saw was a plaid consisting of both diagonals. Yet analysis of the hemodynamic response of primary visual cortex revealed that this region distinguished between the left and the right fringes. In other words, primary visual cortex—but not higher visual regions such as V2—saw the orientation of the invisible stripes, yet that information was inaccessible to the owner of the brain, in agreement with our hypothesis.

The primary visual cortex is the gateway to about three dozen other regions of the cerebral cortex dedicated in some fashion to visual processing. Given its strategic location, it is ironic that the primary visual cortex is not even *necessary* for all forms of visual perception. Brain imaging in dreaming volunteers (no easy feat, given the tight quarters and the noisy banging inside a scanner) suggests that activity in the primary visual cortex is curtailed in the rapid eye movement (REM) phase of sleep, when most dreams occur, compared with non-REM sleep, when little dreaming takes place. Furthermore, patients with damage to their primary visual cortex dream without any concomitant loss of visual content.

None of the other primary sensory cortices—the ones that first receive the associated sensory data streams—mediate consciousness either. Loud noises or painful electric shocks fail to elicit any meaningful reaction in people with massive brain damage that condemns them to living in a vegetative state, without any awareness (I'll discuss these patients in more detail in the next chapter). When their cortex is scanned, only their primary auditory and somatosensory cortical regions show significant activity to these powerful stimuli. What all of this teaches us is that isolated activity in the input nodes is not sufficient for consciousness. More is needed.

Neurons in Higher Regions of Neocortex Are Closely Allied to Consciousness

Masking and continuous flash suppression are not the only stealth techniques that visual psychologists use to deliver a payload to the brain below the radar screen of consciousness. *Binocular rivalry* is another one. In binocular rivalry, a small picture, say of a face, is shown to your left eye, and another photo, say the old imperial flag of Japan (with its rays of light emanating from a central disk) to your right eye. You might think that you see the face superimposed onto the flag. If the illusion is set up correctly, though, you will perceive the face alternating with the flag. Your brain won't let you see two things at the same time in the same place.

At first, you vividly see the face without a hint of the sunburst pattern; after a couple of seconds, a patch of the flag appears somewhere in your field of view, erasing the underlying facial pattern at that location. From this initial seed, the image spreads until the face is entirely gone and only the flag remains. Then the eyes start to shine through. The patchwork of flag–face image resolves itself a few seconds later into a whole face. Then the percept of the flag establishes dominance again. And on and on. The two images move in and out of consciousness in a never-ending dance. You can abort this dance by closing one eye—you'll instantly resolve all ambiguity and perceive the image present to the open eye.

The neurophysiologist Nikos Logothetis and his colleagues at the Max Planck Institute in Tübingen trained monkeys to report their percepts during binocular rivalry. Whereas human volunteers are financially rewarded for their labors—cash works best—thirsty monkeys receive sips of apple juice. The monkeys learned over many months to pull one lever whenever they saw the face, to pull a second lever when they saw the sunburst, and to release both when they saw anything else, such as a quilt of both images. The distribution of dominance times—how long the sunburst or the face was seen at a time—and the way in which changing the contrast of the images affected the monkeys' reports, leaves little doubt that both monkeys and people have qualitatively similar experiences.

Logothetis then lowered fine wires into the monkey's cortex while the trained animal was in the binocular rivalry setup. In the primary visual cortex and nearby regions, he found only a handful of neurons that weakly modulated their response in keeping with the monkey's percept. The majority fired with little regard to the image the animal saw. When the monkey signaled one percept, legions of neurons in the primary visual cortex responded strongly to the suppressed image that the animal was not seeing. This result is fully in line with Francis's and my hypothesis that the primary visual cortex is not accessible to consciousness.

The situation was quite different in a higher-order visual area, the *inferior temporal cortex*. Cells in this region responded only to pictures that the monkey saw (and reported): None of them responded to the invisible image. One such neuron might fire action potentials only when the animal indicated that it was seeing the face. When the monkey pulled the other lever, indicating that it was now seeing the flag, the cell's firing activity would fall off substantially—sometimes to zero—even though the picture of the face that excited the cell a few seconds earlier was still present on the retina. This tightly synchronized *pas de deux* between the waxing and waning periods of the cell's activity and the animal's perceptual report reveals a compelling link between a group of neurons and the content of consciousness.

As I mentioned earlier, Francis's and my speculations about the neural correlates of consciousness center on the establishment of a direct loop between neurons in higher-order sensory regions of the cerebral cortex (in the case of vision, the inferior temporal cortex) and their targets in the prefrontal cortex. If prefrontal neurons reach with their axons back into the inferior temporal cortex, then a reverberatory feedback loop is established and can maintain itself. Spiking activity can then spread to regions that underlie working memory, planning, and, in humans, language. Taken as a whole, this coalition of neurons mediates awareness of the face and some of its attendant properties, such as gaze, expression, gender, age, and so on. If a competing loop comes into being, representing the flag, it suppresses activity in the face loop, and the content of consciousness shifts from the face to the sunburst.

Recently, clinicians recorded the EEG of two classes of severely brain-injured patients, those that remain unconscious and those that recover at least some measure of awareness. They found that the critical difference is the presence or absence of communication between prefrontal regions and temporal, sensory cortical regions in the back. If such feedback is present, consciousness is preserved. If not, it is absent. This is a rather gratifying development.

These are early days. We cannot yet pinpoint which regions of the brain underlie consciousness. But this is barking up the wrong tree—we must resist the hypnotic appeal of hot spots in brain scans with their naïve phrenological interpretation: the perception of faces is computed over here, pain over there, and consciousness just yonder. Consciousness does not arise from regions but from highly networked neurons within and across regions.

One singular feature of the brain that has only become apparent over the past two decades is the astounding heterogeneity of neurons. The approximately 100,000 neurons packed below each square millimeter of cortex, an area about the size of the letter "o" on this page, are highly heterogeneous. They can be distinguished based on their location, the shape and morphology of their dendrites, the architecture of their synapses, their genetic makeup, their electrophysiologic character, and the places to which they send their axons. It is critical to understand how this tremendous diversity of actors, perhaps up to a thousand neuronal cell types that are the basic building blocks of the central nervous system, contributes to the genesis of qualia.

The bottom line is that these physiologic experiments are steadily narrowing the gap between the mind and the brain. Hypotheses can be put forth, tested and rejected or modified. And that is a great boon after millennia of sterile debates.

Attending to Something, yet not Seeing It

What is the relationship between selective attention and consciousness? We seem to become conscious of whatever the attentional spotlight illuminates. When you strain to listen to the distant baying of coyotes over the sound of a campsite conversation, you do so by attending to the sound and become consciously aware of their howls. Because of the intimate relationship between attention and consciousness, many scholars conflate the two processes. Indeed, in the early 1990s, when I came out of the closet to give public seminars on the mind–body problem, some of my colleagues insisted that I replace the incendiary "consciousness" with the more neutral "attention," because the two concepts could not be distinguished and were probably the same thing anyway.

I intuitively believed that the two were different. Attention selects part of the incoming data for further scrutiny and inspection, at the cost of disregarding the nonattended portion. Attention is evolution's answer to information overload; it is a consequence of the fact that no brain can process all incoming information. The optic nerve leaving the eye carries a couple of megabytes per second, very little by the standards of today's wireless networks. That information must not only be shipped to the cortex but also be acted upon. The brain deals with this deluge of data by selecting a small portion for further processing—and these selection mechanisms are distinct from consciousness. Thus, attention has a clear functional role to play different from consciousness' mission.

Two decades later, I am confident that the distinction between attention and consciousness is a valid one. As I just discussed, continuous flash suppression obscures images for minutes at a time. Beneath this cloak of invisibility, the astute experimentalist has sufficient scope to manipulate the viewer's attention. That's what Sheng He did, demonstrating that invisible pictures of naked men drew the attention of women, and invisible pictures of naked women attracted the attention of men. Attention is, after all, selective processing of images. A host of other experiments bear this out. For example, functional brain imaging of primary visual cortex during masking reveals that paying attention to invisible objects enhances the brain's response to them. Conversely, manipulating the visibility of the objects had little consistent effect on the hemodynamic response in V1. Clearly, the brain can attend to objects it doesn't see.

Can the opposite, consciousness bereft of attention, also occur? When you attend to a particular location or object, intently scrutinizing it, the rest of the world is not reduced to a tunnel, with everything outside the focus of attention fading: You are always aware of some aspects of the world surrounding you. You are aware that you are reading a newspaper or driving on a freeway with an overpass coming up.

Gist refers to a compact, high-level summary of a scene—a traffic jam on a freeway, crowds at a sports arena, a person with a gun, and so on. Gist perception does not require attentional processing: When a large photograph is briefly and unexpectedly flashed onto a screen while you're being told to focus on some itsy-bitsy detail at the center of your focus, you still apprehend the essence of the photo. A glimpse lasting only one-twentieth of a second is all it takes. And in that brief time, attentional selection does not play much of a role.

As a teenager, my son would talk to me while continuing to play a fast-paced, shoot-'em-up video game. His mind was obviously not fully occupied by our conversation, leaving sufficient attentional resources and bandwidth for a second, more demanding, and, to him, more important task.

Jochen Braun, a psychophysicist at the University of Magdeburg in Germany, perfected a laboratory version of such dual-processing tasks that measures how much somebody can see outside the spotlight of attention. Braun's idea is to pin down attention by giving volunteers a difficult job at the center of their gaze (such as counting how many Xs appear in a stream of letters) and a secondary task somewhere else on the computer screen. The experiment probes how performance deteriorates when this task is done at the same time as the primary, attention-demanding job.

Braun found that with attention engaged at the center of gaze, observers can distinguish out of the corners of their eyes photos with animals in them—a savanna with a lion, a canopy with a flock of birds, a school of fish—from pictures that had none. Yet they are unable to tell a disk that is bisected into red and green halves from its mirror image, a green-red disk (the way the experiment is set up controls for the reduced acuity in the visual periphery). Subjects can judge the gender of a face presented outside their central vision, or whether the face is famous, but are frustrated by jobs that seem much simpler, such as distinguishing a rotated letter "L" from a rotated "T." Braun's experiments show that at least some visual behaviors can be done in the absence—or, adopting a more careful stance—in the near-absence of selective attention.

In the final analysis, psychological methods are too edentate to fully resolve this issue. Without delicately intervening into the underlying brain circuits the distinction between attention and consciousness will not be fully resolved. This is now rapidly becoming possible in laboratory animals such as mice or monkeys. The ultimate test would be to conditionally turn off—and then on again—the control lines in the brain through which attention acts and to observe which visual behaviors the animal is still capable of. I shall return to such experiments in chapter 9.

The history of any scientific concept—energy, atom, gene, cancer, memory—is one of increased differentiation and sophistication until it can be explained in a quantitative and mechanistic manner at a lower, more elemental level. The two-way dissociations I just discussed, attention without consciousness and consciousness without attention, put paid to the notion that the two are the same. They are not. Much of the existing experimental literature will have to be reexamined in the light of this distinction between attention and consciousness. This won't be easy. The distinction I make clears the decks for a concerted, neurobiological attack on the core problem of identifying the necessary causes of consciousness in the brain.

Chapter 5: In which you learn from neurologists and neurosurgeons that some neurons care a great deal about celebrities, that cutting the cerebral cortex in two does not reduce consciousness by half, that color is leached from the world by the loss of a small cortical region, and that the destruction of a sugar cube–sized chunk of brain stem or thalamic tissue leaves you undead

In other terms, there is in the mind groups of faculties, and in the brain, groups of convolutions; and the facts acquired up to now by science permit us to accept, like I have said elsewhere, that the large regions of the mind correspond to the large regions of the brain. It is in this sense that the principle of localization seems to me, if not rigorously demonstrated, at least extremely probable. But to know with certainty whether each particular faculty has its own seat in a particular convolution, this is a question that seems to me all but insoluble in the current state of science.
—Paul Broca, *Bulletin de la Society Anatomique* (1861)

Historically, the clinic has been the most fecund source of insight about the brain and the mind. The vagaries of nature and of men, with their cars, bullets, and knives, give rise to destruction that, when limited in scope, illuminates the link between structure and function and brings out traits that are barely apparent in good health. Let me tell you about four important lessons that patients and their attending physicians, neurologists, and neurosurgeons have taught us about the neural basis of consciousness.

I frequently receive lengthy, unsolicited cogitations in the mail—densely scribbled manuscripts with the promise of more to come, self-published books, or links to extensive Web pages—concerning the ultimate answers to life and consciousness. My attitude to these outpourings is that unless they respect such hard-won neurologic and scientific knowledge, they are destined for the constantly growing X-file sitting in a dusty corner of my office.

Small Chunks of Gray Matter Mediate Specific Content of Consciousness

Many scholars argue that consciousness is a holistic, *gestalt* property of the brain. They reason that it is so otherworldly that it cannot be caused by any one particular feature of the nervous system. Instead, consciousness can only be ascribed to the brain as a whole. In a technical sense that I shall outline in chapter 8, they are correct: Phenomenal consciousness is a property of an integrated system of causally interacting parts. Yet consciousness also has surprisingly local aspects.

A stroke, a car accident, a virus infection, the controlled trauma of a neurosurgeon's scalpel can all destroy brain matter. In their wake they often leave permanent deficits. What is of great interest to the neuroscientist is when the damage is limited and circumscribed. The fact that losing a particular chunk of nervous tissue turns the world into gray tones and well-known faces into unfamiliar ones indicates that this region must be, at least partially, responsible for generating the sense of color or face identity.

Take the case of patient A.R., studied by Jack Gallant at the University of California at Berkeley. At age fifty-two, A.R. suffered a cerebral artery infarct that briefly blinded him. MRI scans done two years later revealed a pea-sized lesion on the right side of his higher visual centers, beyond the primary visual cortex. When Gallant and his colleagues tested A.R. in the laboratory, they discovered that he had lost color vision. Not everywhere, though, but only in the upper left quadrant of his field of view, exactly where they had expected to find it based on the MRI scan. Remarkably, the man was mostly unaware of the graying of part of his world.

A.R.'s low-level vision and his motion and depth perception were normal. The only other deficit was a partial inability to distinguish forms—he couldn't read text—but this was confined, again, to the upper left quadrant of his field of vision.

A pure loss of color perception is called *achromatopsia*. This is quite different from everyday, hereditary color blindness that mainly afflicts men. Because they miss the gene for one of the color pigments in the eye, these dichromats don't perceive as rich a color palette as normal sighted people (trichromats) with three retinal photopigments. Achromatopsia, in contrast, follows destruction of the color center in the visual cortex. In its wake, all hues are leached from the world. No more glorious violet- and purple-tinted Alpenglow in the setting sun. Instead, the world is experienced in chiaroscuro, like a color television that has reverted to shades of black and white. Notably, colors words and associations with colors (e.g., "red" with "fire truck") remain.

Other bizarre disorders abound. People with face-blindness—doctors call it *prosopagnosia*—have troubles with faces. They are unable to recognize famous or familiar ones. They realize that they are looking at a face, they just don't recognize it. All faces look alike, about as distinguishable as a bunch of pebbles in a riverbed. Even though both faces and rocks have many distinct features, you'll have a hard time distinguishing among these smooth rocks, but you can easily recognize hundreds of faces. The reason is that quite a bit of your brain's circuitry is dedicated to processing faces, but very little to handling the appearance of stones (except perhaps if you are a rock collector or a geologist). Because the neurons that mediate the identity of faces in the higher regions of the cerebral cortex were destroyed or were never there in the first place (quite a few people have innate face-blindness), these individuals don't recognize their spouse in a crowd at the airport. They lack the facile and instant recognition you and I experience whenever we see our loved ones.

Face-blindness leads to social isolation and shyness because the afflicted have difficulty recognizing, let alone naming, the people with whom they are conversing. They adopt coping strategies, focusing on some singular mark, a mole or a prominent nose, a brightly colored shirt, or on the voice. Makeup and changing hairstyles impede recognition, as do groups of people in uniforms.

In severe face-blindness, patients don't even see a face as a face. There is nothing wrong with their visual apparatus: They perceive the distinct elements making up a face—the eyes, nose, ears, and mouth—but they can't synthesize them into the unitary percept of a face. *The Man Who Mistook His Wife for a Hat*, a brilliantly observed collection of case studies by the neurologist Oliver Sacks, takes its title from such an individual who tried to shake hands with a grandfather clock when he mistakenly identified the face of the clock as a human face.

Intriguingly, these individuals may still have autonomic reactions to a familiar face: They have an enhanced galvanic skin response—essentially, they sweat a bit—when looking at well-known politicians, movie stars, co-workers, or family members, compared to looking at pictures of people unknown to them. Yet they insist all the while that they do not recognize them. Thus, the unconscious has its own ways of processing emotionally wrought faces.

The flip side of prosopagnosia is the *Capgras delusion*. The patient with this condition persistently claims that his wife has been replaced by an alien, an impostor who looks, talks, and moves the same way the spouse did but is somehow different. The disorder can be quite circumscribed and the patient otherwise unremarkable. Here, face recognition is intact, but the autonomic reaction of *familiarity* is missing. As the patient lacks that emotional jolt that we all take for granted when we encounter somebody intimate, somebody we are accustomed to, he feels that something is terribly amiss.

Akinetopsia is a rare and devastating motion blindness. The individual with this disorder is banished to a world lit only by strobe lights, like a disco or nightclub. The dancers are clearly illuminated by each flash, but they appear frozen, devoid of motion. Seeing yourself move in this fashion in a mirror is mesmerizing, but I can assure you that the excitement wears off quickly. The motion-blind patient infers that objects have moved by comparing their relative positions in time, but she does not see them move. She can see a car change position, but cannot see it drive toward her. Other aspects of vision, including color, form, and the ability to detect flickering lights, are intact.

Based on careful observations of people with such focal damage, Semir Zeki at University College London coined the term *essential node* for the portion of the brain that is responsible for a particular conscious attribute. One region of the visual cortex contains an essential node for the perception of color; several such regions are involved in face perception and in the sense of visual movement. Parts of the amygdala are essential to the experience of fear. Damage to any one node leads to loss of the associated perceptual attribute, although other conscious attributes remain.

The interpretation of clinical data is not as straightforward as I have described it because the brain, especially the young brain, has tremendous powers of recuperation. Even though an essential node is lost, information may be re-routed and re-expressed, and the individual may slowly regain the lost function.

The take-home message is that small chunks of the cerebral cortex are responsible for specific conscious content. This bit of the cortex endows phenomenal experience with the vividness of faces, that part provides the feeling of novelty, and the one over there mediates the sound of voices. The linkage between cortical location and function is a hallmark of nervous systems. Contrast this with another vital organ, the liver. Like the brain, it weighs about three pounds and has a left and a right lobe. But liver tissue is much less differentiated and more homogeneous than nervous tissue. Liver impairment is proportional to the amount of damage, with little regard to where the destruction is located.

Concept Neurons Encode Homer Simpson and Jennifer Aniston

I graphically recall the first time I fell. I had recently started to climb out of sheer desperation—my son had left home for college and my daughter would depart at the end of the school year. The long-dreaded empty nest had arrived! To do something, anything, to soak up my excess energy and enthusiasm, I took up mountain running and rock climbing.

I was leading a crack climb in Joshua Tree, in the California desert. I well remember the exact spot on that rightward-curving, nearly vertical wall of orange-tan granite with its embedded crystals. Granite can shred a climber's hand, yet it is also his best friend because its uneven texture provides extra friction. With my left foot wedged into the crack and my right foot smearing outside, I reached high, high above my head with my right hand to shove a small piece of protective equipment, called a cam, into the crack. It went in too smoothly, and I was afraid it might pop out equally quickly when weighted. I repositioned the cam, jamming it deeper into the crack. At that point my left foot slipped, and I fell ten to twelve feet to the ground, directly onto my back, next to a sharply pointed rock, a potential calamity that I avoided by dumb luck. The skin was burned off my back and I limped for days, but these minor injuries only enhanced the nimbus of climbing. These events indelibly impressed themselves into my memory.

How can the content of consciousness be so replete with particular and evocative details? There are no pictures inside my skull of me climbing, only a brown-grayish organ with the consistency, size, and shape of overcooked cauliflower. This tofu-like tissue, buffered by blood and cerebrospinal fluid, consists of nerve and glial cells. Neurons and their interconnecting synapses are the atoms of perception, memory, thought, and action. If science is ever to comprehend these processes, it must be able to explain them in terms of the interactions of large coalitions of neurons embedded within a network of unfathomable complexity. Consider an analogy: Chemists have no hope of understanding the makeup of matter at normal temperatures unless they know about the electromagnetic forces that govern the interactions of electrons and ions.

The question posed at the start of the last paragraph is a deep one, so far without any definite answer. But I can tell you about one relevant discovery that I was intimately involved with that does lift the veil a bit.

Epileptic seizures—hypersynchronized, self-maintained neural discharges that can engulf the entire brain—are a common neurologic disorder. In many people, these recurrent and episodic brain spasms are kept in check with drugs that dampen excitation and boost inhibition in the underlying circuits. Medication does not always work, however. When a localized abnormality, such as scar tissue or developmental miswiring, is suspected of triggering the seizure, neurosurgery to remove the offending tissue is called for. Although any procedure that breaches the skull carries with it some risks, this surgery is beneficial for people whose seizures can't be controlled in any other way.

To minimize side effects and the loss of quality of life after surgery, it is vital to pinpoint the exact location in the brain from which the seizures originate; this is done by neuropsychological testing, brain scans, and EEG. If no structural pathologies are apparent from the outside, the neurosurgeon may insert a dozen or so electrodes into the soft tissue of the brain, via small holes drilled through the skull, and leave them there for a week or so. During this time, the patient lives and sleeps in the hospital ward, and the signals from the wires are monitored continuously. When a seizure occurs, epileptologists and neuroradiologists triangulate the origin of the aberrant electrical activity. Subsequent destruction or removal of the offending chunk of tissue reduces the number of seizures—and sometimes eliminates them entirely.

The neurosurgeon and neuroscientist Itzhak Fried at the University of California, Los Angeles School of Medicine is one of the world's masters in this demanding trade, which requires great technical finesse. Brain surgeons share many traits with rock climbers and mountaineers, a set of attitudes and behaviors to which I also aspire. They are geeks—exulting in high technology and precision measurements, but they are also sophisticated and literary. They take a blunt, no-nonsense approach to life and its risks; they know their limits but have supreme confidence in their skills. (You don't want your surgeon to be diffident and hesitant when he is about to drill into your cranium.) And they can focus for hours on the task at hand, to the exclusion of everything else.

Itzhak and his surgical colleagues perfected a variant of epilepsy monitoring in which the electrodes are hollowed out. This permits them to insert tiny wires, thinner than hair, straight into the gray matter. Using appropriate electronics and fancy signal detection algorithms, the bundle of miniaturized electrodes can pick up the faint chattering of a bevy of ten to fifty neurons from the ceaseless background cacophony of the electrical activity of the brain.

Under Itzhak's supervision, a group from my laboratory, Rodrigo Quian Quiroga, Gabriel Kreiman, and Leila Reddy, discovered a remarkable set of neurons in the jungles of the medial temporal lobe. This region, which includes the hippocampus, turns percepts into memories, but it is also the source of many epileptic seizures, which is why Itzhak places electrodes here.

We enlist the help of the patients. While they have nothing better to do than to wait for their seizures, we show them pictures of familiar people, animals, landmark buildings, and objects. We hope that one or more of the photos strike the fancy of some of the monitored neurons, prompting them to fire a burst of action potentials. Most of the time the search turns up empty-handed, although sometimes we come upon neurons that responded to categories of objects, such as animals, or outdoor scenes, or faces in general. But a few neurons are much more discerning. I was thrilled when Gabriel showed me the first such cells. One fired only when the patient was looking at photos of then-President Bill Clinton, but not other famous people, and the other responded exclusively to cartoons of Bart and Homer Simpson.

Despite our considerable initial skepticism about this finding—such startling selectivity at the level of individual nerve cells was unheard of—medial temporal lobe neurons are indeed extremely picky about what excites them. One hippocampal neuron responded only to seven different photos of the movie star Jennifer Aniston but not to pictures of other blonde women or actresses. Another cell in the hippocampus fired only to the actress Halle Berry, including a cartoon of her and her name spelled out. We have found cells that respond to images of Mother Teresa, to cute little animals ("Peter Rabbit cell"), to the images of the dictator Saddam Hussein, and his spoken and written name, and to the Pythagorean theorem, $a^2 + b^2 = c^2$ (this one in the brain of an engineer with mathematics as a hobby).

Itzhak refers to these cells as *concept neurons*. We try not to anthropomorphize them, to avoid the temptation to call them "Jennifer Aniston cells" (the cells don't like it when you do that!). Each cell, together with its sisters—for there are likely thousands of such cells in the medial temporal lobe for any one idea—encodes a concept, such as Jennifer Aniston, no matter whether the patient sees or hears her name, looks at her picture, or imagines her. Think of them as the cellular substrate of the Platonic Ideal of Jennifer Aniston. Whether the actress is sitting or running, whether her hair is up or down, as long as the patient recognizes Jennifer Aniston, those neurons are active.

Nobody is born with cells selective for Jennifer Aniston. Like a sculptor patiently releasing a *Venus de Milo* or *Pieta* out of a block of marble, the learning algorithms of the brain sculpt the synaptic fields in which concept neurons are embedded. Every time you encounter a particular person or object, a similar pattern of spiking neurons is generated in higher-order cortical regions. The networks in the medial temporal lobe recognize such repeating patterns and dedicate specific neurons to them. You have concept neurons that encode family members, pets, friends, co-workers, the politicians you watch on television, your laptop, that painting you adore. We surmise that concept cells represent more abstract but deeply familiar ideas as well, such as everything we associate with the mnemonic 9/11, the number pi, or the idea of God.

Conversely, you do not have concept cells for things you rarely encounter, such as the barista who just handed you a nonfat chai latte tea. But if you were to befriend her, meet her later on in a bar, and let her into your life, the networks in the medial temporal lobe would recognize that the same pattern of spikes occurred repeatedly and would wire up concept cells to represent her.

Many neurons in visual cortex react to a line with a particular orientation, to a patch of gray, or to a generic face with promiscuous exuberance, whereas concept cells in the medial temporal lobe are considerably more restrained. Any one individual or thing evokes activity in only a very small fraction of neurons. This is known as a sparse representation.

Concept cells demonstrate compellingly that the specificity of conscious experience has a direct counterpart at the cellular level. Say you are recalling the iconic scene of Marilyn Monroe standing on a subway grill, keeping the wind from blowing her skirt up. It is commonly assumed that the brain uses a broad population strategy to represent this percept. A few tens of millions of nerve cells fire in one manner when you see Monroe and in a different way when you see Aniston, the Queen of England, or your grandmother. It's always the same population of cells that respond, but in different ways. Yet our discovery makes this unlikely for those concepts or individuals that you are intimately familiar with. Most cells are silent most of the time, the essence of a sparse representation. When Monroe appears, a small minority fires; a different group will be active to Aniston and so on. Any one conscious percept is caused by a coalition of neurons numbering perhaps in the hundreds or thousands rather than in the millions.

More recently, Moran Cerf and others from my lab, together with Itzhak, hooked the signals from several concept cells to an external display to visualize a patient's thoughts. The idea is deceptively simple but fiendishly difficult to implement. It required three years of effort by Moran—a computer security specialist and movie maker turned Caltech graduate student—to pull off this feat.

Let me walk you through one example. Moran recorded from a neuron that fired in response to images of the actor Josh Brolin (whom the patient knew from her favorite movie, *The Goonies*) and from another neuron that fired vicariously in response to the Marilyn Monroe scene I just mentioned. The patient looked at a monitor where these two images were superimposed, with the firing activity of the two cells controlling the extent to which she saw Brolin or Monroe in the hybrid image (via feedback from the patient's brain to the monitor). Whenever the patient focused her thoughts on Brolin, the associated neuron fired more strongly. Moran arranged the feedback such that the more this cell fired relative to the other one, the more visible Brolin became and the more the image of Monroe faded and vice versa. The image on the screen kept changing until only Brolin or only Monroe remained visible and the trial was over. The patient loved it, as she controlled the movie purely with her thoughts. When she focused on Monroe, the associated neuron increased its firing rate, the cell for the competing concept, Brolin, contemporaneously dampened its activity, while the vast majority of neurons remained unaffected.

The way I tell the story here, it sounds like there are two principals, the way the puppeteer Craig occupied the head of actor John Malkovich in the movie *Being John Malkovich*. One is the patient's mind, concentrating on Monroe. The other is the patient's brain—namely, the nerve cells in the medial temporal lobe that up- and down-regulate their activity according to the mind's desire. But both are part of the same person. So who is in control of whom? Who is the puppeteer and who the puppet?

Itzhak's electrodes probe ground zero of the neuronal correlates of consciousness. The patient can quite deliberately and very selectively turn the volume on her medial temporal lobe neurons up and down. But many regions of the brain will be immune from this influence. For instance, you can't will yourself to see the world in shades of gray. This most likely means that you can't consciously downregulate color neurons in your visual cortex. And much as you may sometimes want to, you can't turn off the pain centers in your brain.

All the weirdness of the mind–body nexus is apparent here. The patient doesn't feel an itch every time the Monroe neuron fires; she doesn't think, "Suppression, suppression, suppression," to banish Brolin from the screen. She has absolutely no idea whatsoever what goes on inside her head. Yet the thought of Monroe translates into a particular pattern of neuronal activity. Events in her phenomenal mind find their parallel in her material brain. A mind-quake occurs simultaneously with a brain-quake.

Consciousness Can Be Generated by Either Cortical Hemisphere

The brain, like the rest of the body, has a remarkable degree of bilateral symmetry. It is helpful to think of it as an enlarged walnut. One side is not the exact mirror image of the other, but approximately so. Almost every brain structure has two copies, one on the left and one on the right. The left side of the visual field is represented by the visual cortex in the right hemisphere, whereas the right side is mapped onto the left visual cortex. When you look out at the world, you don't see a fine vertical line running down your field of view; the two hemifields are integrated seamlessly. Philosophers emphasize that experience is unitary. You don't experience two streams of consciousness, one for each side, but only one. And what is true for vision applies with equal force to touch, to hearing, and so on.

The discordance between the two halves of the brain and the one mind was pointed out by Descartes, who was looking for a single organ that reflected the unitary nature of experience. He erroneously assumed that the pineal gland does not have a left and a right half and famously hypothesized that it was the seat of the soul (in modern language, the neural correlate of consciousness). When I mention Descartes' fingering of the pineal gland in class, some students snicker, "How silly." In fact, Descartes was centuries ahead of his time, looking for a relationship between structure and function. He's a breath of fresh air, of modernity and enlightenment in the dusty, moth-eaten atmosphere of the closing years of Medieval scholasticism. Descartes replaced worn-out Aristotelian teleonomic, final causes that really don't explain anything—wood burns because it possesses an inherent form that seeks to burn—by mechanistic ones. Descartes resides, together with Francis Crick and the late neurosurgeon Joseph Bogen, in my personal pantheon. (If the truth be told, the boy reporter Tintin and the detective Sherlock Holmes are also members.)

The *corpus callosum*, the largest white matter structure in the brain, is primarily responsible for this integration. It is a thick bundle of about 200 million axons, each extending from a pyramidal cell on one side of the brain to the other side. These axons, together with some minor wire bundles, tightly coordinate the activities of the two cerebral hemispheres so that they work together effortlessly, giving rise to a single view of the world.

What happens if this bundle of axons is cut? If it were done carefully, without causing damage to other structures, the patient should remain sentient, though his or her consciousness might be cut in two, shrinking to encompass only the left or the right visual field, with the other half invisible. However, this is not what happens!

In certain cases of intractable epileptic seizures, part or all of the corpus callosum is cut to prevent a seizure that originates in one hemisphere from spreading into the other and causing generalized convulsions. This operation, first performed in the early 1940s, alleviates seizures. Remarkably, *split-brain* patients, once they've recovered from the surgery, are inconspicuous in everyday life. They see, hear, and smell as before, they move about, talk, and interact appropriately with other people, and their IQ is unchanged. They have their usual sense of self and report no obvious alteration in their perception of the world— no shrinkage of their visual field, for example. The surgeons who pioneered this operation, such as Joseph Bogen at Loma Linda University in Southern California, were puzzled by this lack of clear symptoms.

Closer inspection of split-brain patients by the biologist Roger Sperry at Caltech, however, revealed a persistent and profound disconnection syndrome. If specific data are given to one hemisphere, that information is not shared with its twin on the other side. Furthermore, only one hemisphere, typically the left one, speaks. That is, if the right hemisphere is lost or silenced by anesthesia, the patient can still talk, which is why the left is called the dominant hemisphere. The right hemisphere by itself has only limited language comprehension and is mute, though it can grunt and sing. So, when engaged in conversation with a split-brain patient, it is the person's left hemisphere that is doing all the talking. He can't name an object presented in the left visual field because that image is processed by his mute right visual cortex. But he can pick out the object from a group of objects on a tray using his left hand, which is controlled by the right motor cortex.

If a key is placed into his right hand, which is under the table, out of sight, the patient will quickly name it. The touch information from his right hand is transmitted to his left hemisphere, where the object is identified and its label relayed to the language center. If the key is placed into the person's left hand, however, he is at a loss to say what it is and rambles on. The right hemisphere might very well know that the object is a key, but it cannot convey this knowledge to the language centers on the left, because the communication links have been cut.

One half of the brain quite literally does not know what the other half is doing, which leads to situations somewhere between tragedy and farce. Victor Mark, a neurologist at the University of North Dakota, videotaped an interview with a split-brain patient. When asked how many seizures she had following her operation, her right hand held up two fingers. Her left hand then reached over and forced the fingers on her right hand down. After trying several times to tally her seizures, she paused and then simultaneously displayed three fingers with her right hand and one with her left. When Mark pointed out this discrepancy, the patient commented that her left hand frequently did things on its own. A fight ensued between the two hands that looked like slapstick comedy. Only when the patient grew so frustrated that she burst into tears was I reminded of her sad situation.

Studies with split-brain patients, work for which Sperry was awarded the Nobel Prize in 1981, teach us that cutting the corpus callosum cleaves the cortico-thalamic complex in two but leaves consciousness intact. Both hemispheres are independently capable of conscious experience, one being much more verbal than the other. Whatever the neural correlates of consciousness, they must exist independently in both hemispheres of the cerebral cortex. Two conscious minds in one skull. I return to this theme in chapter 8.

Consciousness Can Flee Permanently, Leaving a Zombie Behind

As long as you are awake, you are conscious of something—the road ahead, a heavy metal piece by *Rammstein* running incessantly through your mind, or fantasies about sex. It is only during certain meditative practices that one can be conscious without having any specific content, aware without being aware of anything in particular. Even when your body is asleep, you can have vivid experiences in your dreams. In contrast, during deep sleep, anesthesia, fainting, concussion, and coma, there is no experience at all. Not a black screen but *nada*.

When severe injury strikes the brain, consciousness may not return. A car accident, a fall, or a combat wound, a drug or alcohol overdose, a near drowning—any of these can lead to profound unconsciousness. Thanks to rescue helicopters and emergency medical technicians, who quickly deliver the victim to the care of a team of specialized trauma nurses and physicians with their advanced tools and pharmaceutical cornucopia, many can be plucked back from the edge of death. Although this is a blessing for most, it is a curse for a few. They remain alive for years, never recovering consciousness, undead.

Such global disorders of consciousness occur when the parts of the brain responsible for arousal are damaged. Neurons in the thalamus and cerebral cortex can't assemble into the far-flung neuronal coalitions that mediate any one conscious content. Impaired states of consciousness include *coma*, the *vegetative state*, and the *minimally conscious state*. Overall arousal fluctuates from complete absence in coma, to periodic sleep–wake transitions in the vegetative state, and awakenings with purposeful movements in the minimally conscious state, sleep walking, and certain partial epileptic seizures.

In the United States alone, as many as 25,000 patients hover for years in a vegetative state termed *persistent vegetative state* (PVS), with bleak prospects for recovery. What makes the situation almost unbearable is that unlike comatose patients, who exhibit almost no reflexes, patients in this limbo state have daily sleep–wake cycles. When they are "awake," their eyes are open and may move reflexively, as do their limbs on occasion; they may grimace, turn their head, groan. To the naïve bedside observer, these movements and sounds suggest that the patient is awake, desperately trying to communicate with her loved ones. The tragedy of the ruined patient's blank and empty life, drawn out over hopeless decades in hospices and nursing homes, is mirrored and amplified by the love—and the resources—her family expends on her care, always hoping for a miraculous recovery.

You may recall Terri Schiavo in Florida, who lingered for fifteen years in a persistent vegetative state until her medically induced death in 2005. Because of the nasty, public fight between her husband, who advocated discontinuing life support, and her parents, who believed that their daughter had some measure of awareness, the case caused a huge uproar, was litigated up and down the judicial chain, and eventually drew in then-President George W. Bush. Medically, her case was uncontroversial. She had brief episodes of automatisms: head turning, eye movements, and the like, but no reproducible or consistent, purposeful behavior. Her EEG was flat, indicating that her cerebral cortex had shut down. Her condition failed to improve over many years. The autopsy showed that her cortex had shrunk by half, with her visual centers atrophied; so contrary to public reports circulating at the time, she couldn't have seen anything.

I'm going to digress for a bit. Current legislation in the United States draws a sharp distinction between withdrawing medical care and active euthanasia. In the former case, the terminally ill patient dies on his or her own schedule. In the latter, the doctor intervenes via opiates or other drugs that speed up the arrival of death. I understand the historical forces that gave rise to laws prohibiting euthanasia. But allowing a patient, even an unconscious one like Schiavo, to die by withdrawing all liquid or solid sustenance and starving to death seems barbaric to me. In her twelfth year, Trixie, our beloved family dog, was suffering from cardiomyopathy. She stopped eating, her belly filled with water, she vomited frequently, and she had trouble controlling her bowels. My wife and I finally took her for that last visit to the vet. He injected her with a large dose of barbiturate while she was resting trustingly in my arms, licking me gently, until her brave heart stopped. It was quick, it was painless—albeit profoundly sad—and it was the right thing to do. I hope that when my time comes, somebody will render me the same service.

Unfortunately, distinguishing between a patient in a persistent vegetative state, who has regular sleep–wake transitions, and somebody in a minimally conscious state, who can sporadically communicate with people around them, is often difficult. One tool for accomplishing this, a sort of consciousness-meter, will be discussed in chapter 9. Functional brain imaging can be another one.

Adrian Owen, a neurologist at the University of Cambridge in England, placed an unresponsive woman whose brain had been severely damaged in a traffic accident into an MRI scanner. She was read instructions by her mother, asking her to imagine playing tennis or visiting all the rooms in her house. The patient showed no signs of comprehending, let alone responding. Yet the pattern of hemodynamic brain activity was similar to that of healthy volunteers who closed their eyes and imagined similar actions. Such fantasizing is a complex and purposeful mental activity that takes place over minutes: It is unlikely to occur unconsciously. The injured woman, despite her inability to signal with her hands, eyes, or voice, was at least sporadically conscious and able to follow an external command. Most other vegetative state patients who were tested in this manner had no such brain signatures; they appear to be truly not conscious. The MRI scanner could be a lifeline for grievously brain-injured patients because it opens up a means of communication: "If you are in pain, think of playing tennis. If not, imagine walking through your house."

To return to the main theme, it is remarkable that large parts of the cerebral cortex can be destroyed without any overall loss of function after recovery. As I stated above, a person with focal cortical damage has limited deficits. This resilience to damage is especially evident in the frontal lobes. Stimulating them with electrical currents does not yield any twitching limbs—as does stimulation of the primary motor cortex—or flashes of light—as does stimulation of the visual cortex. That's why early neurologists often referred to the frontal lobes as *silent regions*.

The defining feature of classical psychosurgery is the controlled destruction of gray matter in the frontal lobes of the cortex (lobotomy) or the cutting of the axons in the white matter that connect the prefrontal cortex to the thalamus and the basal ganglia (leucotomy). These procedures, infamously done with a modified ice pick inserted through the eye socket, cause personality changes and mental disabilities. They turn "the insane into the idiot" and facilitate custodial care of patients, but they do not cause a global loss of consciousness.

Yet a small, confined injury to subcortical structures located close to the imaginary midline separating the left and right brain *can* render a person unconscious. I think of these midline structures as *enabling factors* for consciousness. They control the degree of brain arousal needed for awareness. If both the left and the right copies of a subcortical region are destroyed, the patient may lose consciousness permanently. (In general, the brain tolerates injury to a structure on one side but is much less resistant to damage to both sides.) One such midline structure is the *reticular activating system*, a heterogeneous collection of nuclei in the upper brain stem and hypothalamus. Nuclei are three-dimensional sets of neurons with their own unique cellular architecture and neurochemical identity. The nuclei in the reticular activating system release modulatory neurotransmitters such as serotonin, norepinephrine, acetylcholine, and dopamine from their axons throughout the forebrain.

Another enabling factor for consciousness is the set of five intralaminar nuclei of the thalamus, also clustered around the midline. These nuclei receive input from brain-stem nuclei and the frontal lobes and send their output throughout the cerebral cortex. Lesions no bigger than a sugar cube in both the left and right intralaminar nuclei cause consciousness to flee, most likely permanently.

A plethora of nuclei in the thalamus and the brain stem keep the forebrain sufficiently aroused for experience to occur. None of these structures, with their distinct chemical signatures, is responsible for generating the content of that experience, but they make experience possible. The endpoint of their efforts is the sixteen billion neurons in the cerebral cortex and their close associates in the thalamus, the amygdala, the claustrum, and the basal ganglia. By controlling the release of a cocktail of neurotransmitters, the intralaminar nuclei and other nuclei in the catacombs of the brain tune synaptic and neuronal activity up or down, enabling the cortico-thalamic complex to form and shape the tightly synchronized coalition of neurons that is at the heart of any one conscious experience.

In summary, local properties of the cortex and its satellite structures mediate the specific content of consciousness, whereas global properties are critical for sustaining consciousness per se. For a coherent coalition of neurons to assemble at all—and for awareness to emerge—the cortico-thalamic complex needs to be suffused with neurotransmitters, chemicals released by the long and winding tentacles of neurons in the deeper and older parts of the brain. Both local and global aspects are critical for consciousness.

But enough neuroanatomy and neurochemistry. Let me now turn to the unconscious.

Chapter 6: In which I defend two propositions that my younger self found nonsense—you are unaware of most of the things that go on in your head, and zombie agents control much of your life, even though you confidently believe that you are in charge

What does man actually know about himself? Is he, indeed, ever able to perceive himself completely, as if laid out in a lighted display case? Does nature not conceal most things from him—even concerning his own body—in order to confine and lock him within a proud, deceptive consciousness, aloof from the coils of the bowels, the rapid flow of the bloodstream, and the intricate quivering of the fibers! She threw away the key.
—Friedrich Nietzsche, *On Truth and Lying in an Extramoral Sense* (1873)

It was only as a mature man that I became mortal. The visceral insight of my end came to me abruptly more than a dozen years ago. I had wasted an entire evening playing an addictive, first-person shooter video game that belonged to my teenage son—running through eerily empty halls, flooded corridors, nightmarishly twisting tunnels, and empty plazas under a foreign sun, emptying my weapons at hordes of aliens pursuing me relentlessly. I went to bed late and, as always, fell asleep easily. I awoke abruptly a few hours later. Knowledge had turned to certainty—I was going to die! Not right there and then, but someday.

I did not have any premonition of accidents about to happen, cancer, or so on—just the sudden, deeply felt realization that my life was going to end. Death had paid an unannounced visit more than a decade earlier when Elisabeth, the identical twin of my daughter Gabriele, was taken from us by crib death (sudden infant death syndrome) at eight weeks. Children should not die before their parents; it profoundly violates the natural order of things. This horrid experience tainted everything in its wake, but strangely, it had not affected my own feelings of mortality. But this nocturnal insight was different. I now understood, deeply understood, that I, too, was going to die. The certainty of death has remained with me, making me wiser but no happier.

My interpretation of this queer event is that all the killing in the video game triggered unconscious thoughts about the annihilation of the self. These processes produced sufficient anxiety that my cortico-thalamic complex woke up on its own, without any external trigger. At that point, self-consciousness lit up and was confronted with its mortality. This singular, yet universal experience vividly brought home the insight that much of what goes on in my head is not accessible to me. Somewhere in the brain, my body is monitored; love, joy, and fear are born; thoughts arise, are mulled over, and discarded; plans are made; and memories are laid down. The conscious me, Christof, is oblivious to all of this furious activity.

Suppressing knowledge of the certain doom that awaits us all must have been a major factor in the evolution of what Freud calls *defense mechanisms* (are we the only animals with them? can a chimp suppress or repress?). These are processes by which the brain removes negative feelings, anxiety, guilt, unbidden thoughts, and so on from consciousness. Without such cleansing mechanisms, early humans might have become too transfixed by their ultimate fate to successfully dominate their niche. Perhaps clinical depression amounts to a loss of such defense mechanisms.

But with the right trigger, the unconscious can manifest itself dramatically. Whenever I'm in Boston, I try to visit Elisabeth's grave. A few years after my late-night revelation, I made the pilgrimage to St. Joseph Cemetery. As I was walking past row after row of headstones in a gentle spring rain, all by myself, I noticed from afar something odd about her gravestone. Up close, I was stunned to see a small terra cotta angel with broken wings on top of the granite block bearing Elisabeth's name. The helpless figurine at the site where my daughter lay buried evoked an immediate and almost unbearable sensation of grief and loss. I sank down onto my knees and wept in the rain. I called my wife and she calmed me down from afar, but I remained shaken for the rest of the day. I never did figure out how the mutilated statue got there. What I learned that day is that a symbol, in the right context, can abruptly release long-dormant memories and emotions.

In my university days, two close friends underwent primal therapy, a form of psychotherapy popularized by the Beatle John Lennon. Ever the sensitive guy, I made great fun of their insistence that repressed memories and instinctual desires and needs were influencing them in ways unbeknownst to them. I vehemently asserted that I was in complete charge of what went on inside my head, thank you very much. I forcefully denied that either the Freudian unconscious or traumatic memories that I didn't even know I had, including the pain of birth—yes, that's part and parcel of primal therapy (you won't be surprised to hear that it originated in Southern California)—were influencing my behavior.

Three decades later, I am more circumspect. I now understand that the actions of the sovereign "I" are determined by habits, instincts, and impulses that largely bypass conscious inspection. The way my nervous system weaves my body effortlessly through a busy shopper lane filled with pedestrians rushing about, how it deciphers the sound patterns that enter my ears and turns them into a question somebody is posing, how my incoherent thoughts come tumbling out in reasonable order from my larynx and mouth as speech, why I can't resist buying a garish-colored shirt of electric violet or imperial purple—all of this is beyond the ken of my conscious self. This lack of awareness extends to the highest regions of the mind.

If you have lived through emotionally charged times, you are well acquainted with the strong crosscurrents of resentment and anger, of fear and desperation, of hope, grief, and passion that are your daily companions. At times, the emotional maelstrom threatens your mental stability. Spelunking the caverns of your own subterranean desires, dreams, and motivations, rendering them conscious, and thereby, maybe, making them comprehensible is very difficult. Psychoanalysis and other inference methods are imperfect; they create a new fiction, a different narrative based on intuitive, folk-psychological notions about why people do the things they do. The talking cure may never unearth the actual reasons why a relationship broke asunder. These remain consigned to the dark cellars of the brain, where consciousness does not cast its prying light.

None of this is new. The submental, the nonconscious or unconscious—which I define as any processing that does not directly give rise to an experience—has been a topic of scholarly interest since the latter part of the nineteenth century. Friedrich Nietzsche was the first major Western thinker to explore the darker recesses of humanity's unconscious desire to dominate others and acquire power over them, frequently disguised as compassion. Within the medico-literary tradition, Freud argued that childhood experiences, particularly those of a sexual or traumatic nature, profoundly determine adult behavior, without their influences being recognized. These Freudian concepts have slipped into everyday language and are only slowly being replaced by more brain-based ones.

Let me move from the anecdotal, autobiographical to the more objective realm of science. I won't be discussing case studies of neurotic, upper-class patients lying on a couch and talking incessantly about themselves at a rate of $200 an hour. Instead, I will relate experiments carried out on groups of college kids paid $15 per hour to participate. The overwhelming conclusion from such findings is a humbling one: Your actions are profoundly shaped by unconscious processes to which you are not privy.

Zombies in the Brain

Neurologic and psychological sleuthing has uncovered a menagerie of specialized sensory-motor processes. Hitched to sensors—eyes, ears, the equilibrium organ—these servomechanisms control the eyes, neck, trunk, arms, hands, fingers, legs, and feet, and subserve shaving, showering, and getting dressed in the morning; driving to work, typing on a computer keyboard, and text messaging on your phone; playing a basketball game; washing dishes in the evening; and on and on. Francis Crick and I called these unconscious mechanisms zombie agents. Collectively, this zombie army manages the fluid and rapid interplay of muscles and nerves that is at the heart of all skills and that makes up a lived life.

Zombie agents resemble *reflexes*—blinking, coughing, jerking your hand away from a hot stove, or being startled by a sudden loud noise. Classical reflexes are automatic, fast, and depend on circuits in the spinal cord or in the brain stem. Zombie behaviors can be thought of as more flexible and adaptive reflexes that involve the forebrain.

Zombie agents carry out routine missions below the radar screen of consciousness. You can become conscious of the action of a zombie agent, but only after the fact. I was recently trail running when something made me look down. My right leg instantly lengthened its stride, for my brain had detected a rattlesnake sunning itself on the stony path where I was about to put my foot. Before I had consciously seen the reptile and experienced the attendant adrenaline rush—and before it could give its ominous warning rattle—I had avoided stepping on it and sped past. If I had depended on the conscious feeling of fear to control my legs, I would have trod on the snake.

Marc Jeannerod from the Institut des Sciences Cognitives in Bron, France, is an expert in the neuropsychology of action. His experiments, *sans* snakes, concluded that action can indeed be faster than thought, with the onset of corrective motor action preceding conscious perception by about a quarter of a second. To put this into perspective, consider a world-class sprinter running one hundred meters in ten seconds. By the time he consciously hears the pistol, the runner is already several strides out of the starting block.

Unconscious agents come into being by dint of training. Repeating the same sequence over and over reinforces the individual components until they smoothly and automatically interlink. The more you train, the more effortless and synchronized the whole becomes. Training gives athletes and warriors the fraction-of-a-second advantage that spells the difference between victory and defeat—between life and death.

Take saccades, the rapid eye movements with which you constantly scan your environment. You move your eyes hither and thither three to four times each second, 100,000 times a day—about as often as your heart beats—yet you are rarely, if ever, aware of them. You can consciously control your eyes, such as when you avert your gaze from the ugly split lip on the guy next to you or when you avoid catching the eye of a panhandler, but those are exceptions.

The psychologist Bruce Bridgman at the University of California at Santa Cruz and others demonstrated that your eyes see details your mind is unaware of. In one experiment, a volunteer sat in the dark, fixating on a single light-emitting diode. When the light was turned off and turned on again in a peripheral location, the volunteer quickly shifted her gaze to the new location. On occasion, the experimentalist "cheated" and moved the light for a second time while the volunteer's eyes were already on route, executing the saccade. Her eyes didn't miss a beat, landing right on the light's new position, even though it had shifted. The volunteer herself was unaware of the second shift in the light's position because vision is partially shut down during saccades. (That's why you can't see your eyes move: try looking in a bathroom mirror while shifting your eyes back and forth.) Indeed, when the light was shifted a bit inward or outward, the volunteer couldn't guess which way the light had been moved, even though her eyes still landed right on target.

The saccadic system is exquisitely sensitive to where it has to direct the eyes. Given its high degree of specialization, there is little need to involve consciousness in its stereotyped actions. If you needed to be aware of and plan every eye movement, you couldn't do much else besides. Imagine trying to pursue a train of thought if you constantly had to think, "Now shift left, now down, now over there, then down here again!," while having to program the dozen eye muscles appropriately. Why clog up experience with these banalities if they can be contracted out to specialists?

Zombie agents operate in the here and now. They don't plan for the future. When you reach out to pick up a cup of hot tea, swerve on your bike to avoid a car that is abruptly changing lanes, return a tennis volley, or swiftly type on a keyboard, you need to act now, not a few seconds from now.

Acquiring a new skill, such as sailing or mountaineering, takes a great deal of physical and mental discipline. In rock climbing, you learn where to place your hands, feet, and body to smear, stem, lie back, lock off the wrist or the fingers in a crack. You pay attention to the flakes and grooves that turn a vertical granite cliff into a climbable wall with holds. A sequence of distinct sensory-motor routines needs to be stitched together, assembled, and compiled into an elaborate motor program. Only after hundreds of hours of dedicated training does execution become automatic, turning into what is colloquially referred to as *muscle memory*. Constant repetition recruits an army of zombie agents that makes the skill effortless, the motion of your body fast, without wasted effort. You never give a thought to the minutiae of your actions, yet they require split-second coordination and a marvelous merging of muscle and nerve.

The shift from attention-demanding and awareness-hogging action to automatic and unconscious action is accomplished by a shift of neuronal resources from the prefrontal cortex to the basal ganglia and cerebellum.

Paradoxically, once learning has occurred, paying attention to details can disrupt skilled performance. Focusing on any one aspect of a highly trained skill, say the exact moment when the right inner side of the foot touches the soccer ball during dribbling, slows performance or makes it more error-prone. When performing a well-practiced piece of music that you haven't played in a while, it is best to let your "fingers do the playing." Becoming aware of the fluidity of your actions or thinking about the individual motifs and sequences of notes can lead you astray. Indeed, compliment your tennis opponent on the optimal posture of his back-hand, and the next time he returns the ball he'll pay attention to his "perfect" posture and the ball will go astray.

The psychologists Gordon Logan and Matthew Crump of Vanderbilt University established this paradox for keyboard typing, an essential skill for our time. Volunteers had to type words presented on a screen at their usual typing speed under three conditions. To establish a baseline, they typed the words normally, with both hands. In left trials, they had to suppress their right hands, typing only letters traditionally assigned to the left hand on the keyboard. In right trials, the right hand was used to type only its letters. For example, if the word "army" came up on the monitor in a left trial, the subject had to type "a" and "r" but not "m" or "y," which are supposed to be typed by the right hand. Inhibiting these automatic manual responses was difficult and slowed the volunteers' typing speed markedly, even though only half as many letters had to be typed. Yet, if the forbidden letters were outlined in green and the permissible letters in red, subjects could easily stop their left or right hand from typing, and in good time, too.

Which hand types which letter is a detail that is relegated to lower-level motor routines. Rendering that knowledge conscious in order to suppress action takes effort. However, if the input that the typing zombie reads is already marked as "do not type," there is no need to dredge up this information, and the hand can be stopped in time to avoid slowing overall typing speed. To combat such *choking under pressure*, coaches and training manuals recommend emptying your mind of everything. This frees up your inner zombie, to considerable benefit.

The same thought is expressed in *Zen in the Art of Archery*, a gem of contemplative literature. Toward the end of this book, the author, Eugen Herrigel, has this to say about sword fighting:

The pupil must develop a new sense or, more accurately, a new alertness of all of his senses, which will enable him to avoid dangerous thrusts as though he could feel them coming. Once he has mastered the art of evasion, he no longer needs to watch with undivided attention the movements of his opponent, or even of several opponents at once. Rather, he sees and feels what is going to happen, and at that same moment he has already avoided its effect without there being "a hair's breadth" between perceiving and avoiding. This, then, is what counts: a lightning reaction which has no further need of conscious observation. In this respect at least the pupil makes himself independent of all conscious purpose. And that is a great gain.

The neuropsychologists Melvyn Goodale and David Milner studied a patient who lost much of her sight in a near-fatal incident of carbon monoxide poisoning. Diagnosed with visual agnosia, she can't recognize objects or see their shape or orientation (although she can see colors and textures). The accident deprived parts of her visual cortex of oxygen, killing the neurons there. As a consequence, she can't tell a horizontal line—like the slot in a mail box—from a vertical one. They look the same to her. Yet, if she has to reach out and insert a letter into the slot, she does so without hesitation, regardless of its slant—horizontal, vertical, or in between. Although she is unaware of the slot's orientation, her visual-motor system has access to this information and smoothly guides her hand without groping. Likewise, when the woman has to grasp objects placed in front of her, she does so with considerable accuracy and confidence, even though she can't say whether she is holding a thin glass flute or a big mug. She lost the part of her visual cortex responsible for recognizing objects, but the regions that use retinal input to shape and guide her hand and fingers remain functional.

Earlier clinical work, fortified by experiments in monkeys, gave birth to the idea of two distinct visual processing streams. Both originate in the primary visual cortex but then divide to innervate different regions of the cortex dedicated to higher visual and cognitive functions. One flows through V2 and V3 into the inferior temporal cortex and the fusiform gyrus. This is the *perceptual*, *ventral*, or *what pathway*. The other, the *action*, *dorsal*, or *where pathway*, delivers data to the visual-motor regions in the posterior parietal cortex. Based on a careful analysis of the visual-motor skills and deficits of their singular patient, Milner and Goodale inferred that her *what* pathway, mediating conscious object vision, was destroyed by her near suffocation, whereas her *where* pathway, guiding her hand and fingers, remained largely intact. Its purpose is not endless rumination, the prerogative of certain prefrontal structures, but direct action. A well-functioning brain integrates both pathways tightly, with many cross-connections; the owner of the brain doesn't even realize that the seamless interface he or she experiences is made up of two or more streams of information.

The Social Unconscious

We are intensely social animals. The popularity of celebrity and gossip shows, magazines, and Web sites full of the latest blather, photos, and buzz about parties, affairs, back-stabbings and conspiracies, babies born out of wedlock, and so on is just one, rather exuberant expression of people's endless curiosity about each other's doings. If you think humanity is high-minded, just check out Google's *Zeitgeist* archives for the top ten search terms. Movie and pop stars, bands, top athletes, and current political events are the perennials, with nary a scientist nor scientific discovery among them (and this is after the most popular search terms, those that relate to sex, have already been screened out).

No one is an island. Even the recluse defines himself by his relationships with others, if not in real life, then through books, movies, or television.

You might believe as fervently as I did as a young man that your conscious intentions and your deliberate choices control your interactions
with family members, friends, and strangers. Yet decades of social psychology research have clearly shown otherwise. Your interactions are
largely governed by forces beyond your ken, by unconscious desires,
motivations, and fears.

William James, the father of American psychology and brother of the
novelist Henry James, held that the act of thinking about or observing a
behavior increases the tendency to engage in that behavior. This principle of *ideomotor action* is a natural consequence of the partial overlap
of the representations of perception and action in the cerebral cortex.
The discovery of so-called mirror neurons in the brain of monkeys con-
firms the notion that perception of an action is closely allied to carrying
out that action.

Looking at somebody eating will activate, albeit weakly, the regions in
your brain that are triggered when you eat. Seeing somebody embarrass
himself, you cringe, because you feel a little bit embarrassed, too. When
somebody smiles at you, you feel a little bit better. When you feel
favorably disposed toward someone, you imitate that person's actions
and words. Watch this the next time you're meeting a friend in a café.
Both of you might lean with the same elbow on the table and incline
your head the same way. When you whisper, your friend follows suit.
When she scratches her head, you do, too. When you yawn, she does
likewise. Reciprocity of action also helps when ingratiating yourself to
somebody.

An inexhaustible multiplicity of factors influences your day-to-day
encounters with people. Their age, gender, ethnicity, dress, bearing, and
emotional expression will imprint themselves on your mind and will
guide the way you approach, talk to, and judge them. And all of this
escapes conscious scrutiny; hence the importance of "first impressions."

Some people overtly hold strong, usually negative opinions about
specific groups: "liberals hate the country," "Christians are anti-science
zealots," "black men are aggressive," "old people are boring," and so on.
Such bigots will make choices based on their prejudices. But even if you
try hard to avoid stereotyping, you still harbor unconscious biases and
predilections. You can't escape being a child of your culture and upbringing, soaking up implicit judgments from fairy tales and myths, from
books, movies, and games, from your parents, playmates, teachers, and
contemporaries. If you don't believe me, just take an *implicit association
test* (I recommend the Harvard one, which you can find online), in which

you answer a bunch of questions as quickly as you can. The test measures, in an indirect, oblique manner that is resistant to manipulation or lying, the extent of your bias for or against a particular religion, gender, sexual orientation, or ethnic group.

Unconscious biases can be more pernicious than conscious ones, for two reasons.

First, unconscious dispositions are widespread and automatic, becoming activated in the presence of their triggers. They can be very powerful and shared by entire communities.

Consider the surprise attacks on the U.S. Pacific Fleet stationed in Pearl Harbor by the Imperial Japanese Navy on December 7, 1941, and on the World Trade Center and the Pentagon by al-Qaeda on September 11, 2001. Both were colossal failures of intelligence by large organizations dedicated to defending the country from such debacles. Scholarly and journalistic sleuthing uncovered reams of information that pointed to the impeding strikes days, weeks, and months ahead of time. In the case of September 11, intelligence personnel had warned the administration forty times of the threat posed by Osama bin Laden; all in vain.

Why? Countless government panels and books came to similar conclusions. There was incompetence at many levels. Yet more insidious, much more widespread than individual failings to heed warnings, were the explicit and implicit attitudes of racial arrogance and cultural condescension in the minds of the people who could have made a difference. Admiral Kimmel, the officer in charge of the Pacific Fleet, made it perfectly clear in an unguarded moment during one of many congressional investigations: "I never thought those little yellow sons-of-bitches could pull off such an attack so far from Japan." More than fifty years later, Deputy Secretary of Defense Wolfowitz held his opponents in equal disregard, dismissing bin Laden as "this little terrorist in Afghanistan." Widespread stereotyping blinded the people in power: "How can uneducated people living in caves with towels on their heads threaten us, the most powerful nation on the planet?" The events that led to the financial meltdown of Lehman Brothers and that almost crashed the markets in September 2008 are another example of such pathology of thought. Here, it was the widely held belief that investment risk was under control and could be leveraged away by suitable financial instruments that led to a global recession.

Unconscious dispositions can be deliberately shaped for political or economic gain. For example, the inanities and stupidities that flood the "news" months and months before any major election in media-saturated democracies are supremely annoying, but they have an effect. We are so immersed in the products of the advertisement industry that we cease to note them. Yet there is a reason that the industry spent approximately half a trillion dollars worldwide in 2010 to influence the buying decisions of consumers. It works!

Second, whereas laws and public awareness campaigns can root out overt discrimination, unconscious prejudices are much more difficult to counteract. How can you change something when you don't even realize that there is something that needs changing?

If you're unconvinced, consider the experiments of John Bargh at Yale University, which reveal the strength of social *priming*. In priming, pictures, sounds, or words influence the processing of subsequent stimuli. I'll demonstrate priming if you'll play along and call out the color of objects I will name. Don't be embarrassed. Try it by naming the color aloud, and you'll be surprised. Here goes:

What is the color of blank paper? What is the color of a wedding dress? What is the color of snow? What is the color of egg shells?

Now, without thinking about it, blurt out the answer to the question, "What do cows drink for breakfast?"

If you're like most people, you'll think or say "milk." Only after a while will you realize that this is complete nonsense. The repeated invocation of "white" triggered, or primed, neuronal and, probably most critically, synaptic activity associated with other white things. If you're then asked to name a liquid linked to both cows and breakfast, "milk" will automatically come to mind.

Bargh wanted to prime college students for the concepts "rude" and "polite." He did this by asking them to generate sentences from a list of words, ostensibly to test their language skill. One group had words such as *bold*, *rude*, *disturb*, *intrude*, *brazen*, and *impolitely*, whereas a second group used words conveying the opposite, such as *respect*, *patiently*, *yield*, and *courteous*. Once done, the subjects were told to seek out the experimentalist in the hallway for a second test. But the experimentalist pretended to be busy talking to a confederate and the subject had to wait.

Bargh and his colleagues secretly timed how long it took their primed volunteers to interrupt this conversation. The students who had been primed with polite words were amazingly patient, waiting more than nine minutes before interrupting, whereas the rude group lasted just over five minutes before butting in. None of the participants had any suspicion that the word test influenced how long they waited. The conclusion of this experiment, that the words you hear or use shape your behavior, would not be news to my grandmother, who always preached that tipping, bringing small gifts, and being polite pays off in unknown ways.

A variant of this technique measures prejudice associated with older people. Bargh had volunteers work with words that trigger stereotypes of the elderly, such as *old, lonely, forgetful, retired, wrinkled, ancient, helpless,* and *Florida*, whereas a control group had to form sentences with neutral terms. His ingenious means of tapping into elder bias was to covertly time how long the volunteer took to walk from the test site to the elevator, a distance of about ten meters. The students primed for "elderly" took 8.28 seconds, about 1 second longer than the group that was exposed to words not associated with aging. This is a small but real effect. Students had no intimation that the words impelled them to walk more slowly. If the innocent reading of words not relating to you in any direct way can slow you down, how much more strongly will you react if a friend or your spouse tells you that you are getting old? So mince your words.

Self-help movements insist that having a positive and optimistic attitude makes a difference. Although thinking good thoughts will not cure cancer, it will shape your behavior. And that is one reason why I live where I do and do what I do—Americans in general, and the self-selected group of immigrants to the West Coast in particular, believe that with enough desire, sweat, dedication, and the judicious application of technology, almost anything is possible. I share this can-do attitude. To fail when one has given everything one has is honorable; not to try because one is afraid of failure is a major character flaw.

One striking aspect of unconscious processing is that its existence is vehemently denied by so many, including my younger self. Such instinctual, defensive reactions are particularly strong in academics, who think of themselves as more objective, balanced, and fair-minded than everybody else. As a group, professors go out of their way to compensate for gender and racial biases in hiring and mentoring. But when it comes to politics and religion, academics can be remarkably intolerant, attributing to conservatives or devout believers opinions held only by people on the fringes. Their barely contained disdain for most religions, especially Christianity, is so prevalent that many college students feel uncomfortable expressing any religious sentiments at all.

You may object to the importance of the unconscious for two reasons. First, because accepting it implies loss of control. If you're not making the decisions around here, then who is? Your parents? The media, whose products you consume so avidly? Your friends and peers? Second, because you are unaware of unconscious biases (by definition), you don't know you have them. You won't recall an instance in which you clandestinely judged somebody based on their skin color, gender, or age. When somebody points out such a case, you will come up with many vaguely plausible reasons for why you judged the person the way you did—but the thought that you discriminated against the person won't occur to you. It's uncanny, but that's how the mind works.

To underscore this, let me tell you about *choice blindness*. More than one hundred students at the University of Lund in Sweden had to compare two head shots of young women. The experimentalist held each photo in his hand, side by side, and the students had to decide within a few seconds which looked more attractive and point to her. The photos were then briefly removed from sight. Immediately afterward, the students were shown the face they found prettier again and were asked to explain their choice. In some trials, the experimentalist exchanged the pictures using a sleight-of-hand before asking the student to explain his or her choice. Even though the women depicted in the two photos were quite different, most people were fooled. Only one in four students realized that the photos had been switched and that the one they were looking at was not the one they had just picked (in these cases, the experiment was immediately broken off). The other students blithely justified their "choice," even though it contradicted what they had decided only a few seconds earlier: "She's radiant. I would rather have approached her at a bar than the other one. I like earrings," even though the woman they had selected looked solemn and had no earrings.

Choice blindness is relevant not only to dating but to life in general. You frequently have little idea why you do the things you find yourself doing. Yet the urge to explain is so strong that you make up a story on the spot, justifying your choice, confabulating without realizing it.

Some psychologists take the idea of a hidden yet powerful brain with vast processing resources at its disposal even further. They argue that unconscious processing is actually superior to conscious thinking when you must make decisions that require balancing many competing factors. Consider the choice of what apartment to lease. It depends on many variables: the monthly rent, date of availability, size, location, duration of contract, state of the rooms, and so on. According to *unconscious thought theory*, after you've acquired all of the relevant housing information you shouldn't try to introspect but you should distract yourself—by solving a crossword puzzle, say—before deciding which place to rent. Don't sweat the problem. Think of something else, and your invisible brain will solve the problem for you.

Such proposals have drawn a great deal of public scrutiny with their promise of accessing the vast powers of the subliminal mind. Yet many of these experiments are quite "soft," with low statistical significance and no compelling controls. Such flaws are inherent in studying people — their genetics, environment, diet, physical activity, and so on are very difficult to properly control for. A more cautious reading of the data is that forming a rapid impression and coming to a conscious decision can be better than endlessly second-guessing that first-glance assessment. Make a decision, trust yourself, and stick with it.

I am likewise skeptical about the claimed superiority of the unconscious, on both methodological and theoretical grounds. As noted above, history is replete with instances where the widespread adoption of unconscious biases had disastrous consequences. Real-life decisions will always involve a mixture of conscious and unconscious processes, with some decisions relying more on one or the other. I have yet to see convincing evidence that propositional reasoning of the "if–then" variety, complex symbolic manipulation, or dealing with unplanned contingencies can be successfully tackled without conscious, deliberate, time-consuming, eyebrow-furrowing thought. Otherwise, we would all be Einsteins. This conclusion is also in line with millennia of traditional teaching calling for self-examination and rational, calm cerebration before making any important decision.

What Does the Pervasive Influence of the Unconscious Imply for My Quest?

Upon awakening, your consciousness takes a few seconds to boot up. Once you orient yourself, it provides a stable interface with a dizzyingly rich world, with no freezing or annoying blue-screen-of-death that requires your brain to reboot. Like any good interface, the real work takes place underneath the surface. These are the variegated processes that lead from the capture of reflected light beams by your eyes to the percept of a beautiful woman.

Understanding the importance of the unconscious is an ongoing challenge to psychology and neuroscience. It is also necessary to make sense of your life. For without self-examination, without understanding that your actions are not just the outcome of deliberate, conscious choices, you can't better yourself. None of the data reviewed here suggests that the unconscious has vast, hitherto untapped powers that can be harnessed to solve your love, family, money, or career problems. Those can only be addressed by thoughtful and disciplined acts and habits cultivated over years, a boring message that few want to hear.

My quest is to understand consciousness rather than its absence. Unconscious processing is far less mysterious than consciousness; it is, after all, what computers do. How consciousness enters the world is puzzling. Nonetheless, the widespread existence of zombie behavior and unconscious desires and fears is important to my quest, for three reasons.

First, it raises the concern that I may be putting the cart before the horse. If the domain of the submental, the nonconscious, is so all-encompassing, most of the brain and its activity may not relate to consciousness at all.

This is true! I've been at pains to point out that the neural correlates of visual consciousness are not in the spinal cord, the cerebellum, the retina, or the primary visual cortex. I suspect that the majority of neurons in the higher visual and prefrontal regions of the cortex are supporting cast. Perhaps only a thin neural lace of long-range pyramidal neurons that reciprocally connect the front with the back is responsible for conscious content. If pressed for a number, I would surmise that only a small percentage of neuronal activity at any given time is directly involved in constructing a conscious percept. The vast majority of the ceaseless neuronal activity that is the hallmark of a healthy, wakeful brain plays a subordinate role to consciousness.

Lest you think this implausible, consider the following analogy between the mechanisms underlying consciousness and those underpinning heredity (like any analogy, this one is imperfect). The molecular processes that pass information from one cell to its descendants, including replication, transcription, and translation, involve hundreds of fancy biochemical gadgets—DNA, tRNA, mRNA, ribosomes, scaffolding, and centrosomes, to name but a few. Yet the detailed instructions, the blueprint, for assembling a cell are found in a single, double-stranded molecule of DNA—one very long and stable molecule out of millions making up each cell. And a single spelling error in its sequence can have serious consequences down the line. The mechanisms underlying consciousness are likely to be equally specific. Knocking out any one member of the coalition of cortico-thalamic neurons may lead to a minute perturbation of the associated conscious percept or thought.

Second, zombie agents make life more difficult for consciousness researchers because they force us to separate behavior from consciousness: purposeful, routine, rapid actions do not by themselves imply

sentience. Just because a gravely injured person moves her eyes as somebody enters the ward does not mean that she has any situational awareness. The same holds for premature babies, dogs, mice, and flies.
Stereotyped behavior is no guarantee of subjective states. More is needed
for an organism to be certified as having phenomenal experiences.

Third, the widespread existence of zombie routines throws the question of neural correlates of consciousness into bold relief. Where is the
difference that makes the difference? Is it simply a question of the right
region of the cerebral cortex being active, as Milner and Goodale claim?
Dorsal regions for unconscious actions and ventral regions for conscious
vision? Or can the same circuits be involved in both, depending on the
processing mode?

Francis and I argued that short-lived neural activity, which quickly
leaves the retina and travels rapidly through visual-motor regions of the
cortex and on to motor neurons, is insufficient for consciousness. That
requires a single coalition of cortico-thalamic neurons to establish dominance and maintain itself for a while, something more akin to a standing
wave in physics. I shall pick up the theme of engineering mice to prevent
the formation of a dominant coalition in chapter 9.

Let me now broach a key facet of the mind–body problem that I
deliberately avoided in *The Quest for Consciousness* (2004)—namely, the
question of how much freedom the brain has in its actions. Free will is
the philosophical theme *par excellence*. Its roots stretch back to antiquity.
It is a topic that, sooner or later, confronts each one of us. Surprisingly,
a key aspect of this problem is reducible to a question of perceptual
consciousness. To me, this constitutes a major advance in one of the most
disputatious problems in metaphysics.

Chapter 7: In which I throw caution to the wind, bring up free will, *Der Ring des Nibelungen*, and what physics says about determinism, explain the impoverished ability of your mind to choose, show that your will lags behind your brain's decision, and that freedom is just another word for feeling

You see, there is only one constant. One universal. It is the only real truth. Causality. Action, reaction. Cause and effect.
—The Merovingian in *The Matrix Reloaded* (2003)

In a remote corner of the universe, on a small blue planet gravitating around a humdrum sun in the unfashionable, outer districts of the Milky Way, organisms arose from the primordial mud and ooze in an epic struggle for survival that spanned eons. Despite all evidence to the contrary, these bipedal creatures thought of themselves as extraordinarily privileged, as occupying a unique place in a cosmos of a trillion trillion stars. Conceited as they were, they even believed that they, and only they, could escape the iron law of cause and effect that governs everything. They could do this by virtue of something they called free will, which allowed them to do things without any material reason.

Can you truly act freely? Can you do and say things that are not a direct consequence of your predispositions and your circumstances? Did you choose to read this book of your own accord? It felt like you voluntarily decided to browse through its pages in the face of competing interests, say eating lunch or texting a friend. But is that the whole story? Were there no external causes that influenced you—a reading assignment for a class or a friend praising its fluid style? You might argue that these causes were insufficient, that something else had to intervene: your will. Yet the doctrine of predestination, and its secular cousin, determinism, holds that you could not have acted in any other way. You had no true choice in the matter. You are a lifelong indentured servant to an absolute tyrant. You never had the option of lunch but were destined from the beginning of time to read my book.

The question of free will is no mere philosophical banter; it engages people in a way that few other metaphysical questions do. It is the bedrock of society's notions of responsibility, of praise and blame, of being judged for something you did, whether good or bad. Ultimately, it is about the degree of control you exert over your life.

You are living with a loving and lovely spouse. A chance meeting with a stranger, lasting only a few hours, turns this life utterly upside down. You question everything and move out. You talk for hours on the phone, you share your innermost secrets, you start an *affaire de coeur*. You are in love, a heady and powerful elixir of emotions with many of the hallmarks of obsessive-compulsiveness. You realize perfectly well that this is all wrong from an ethical point of view; it will wreak havoc with many lives, with no guarantee of a happy and productive future. Yet something in you yearns for change.

Such gut-churning choices confront you with the essential question of how much say you really have in the matter. Are you not simply acting out the dictates of evolution, the ancient dance of your DNA that seeks out new avenues to propagate itself? Do your hormones, do your loins leave you any freedom? You certainly feel that you could, in principle, break off the affair and return home. But despite many attempts, you somehow never manage to do so. You knowingly sail into the teeth of a perfect storm, wrecking the marital ship that carried you safely for so many years.

Free will is a scholarly minefield. Arcane arguments have been advanced for and against any conceivable position. In my thoughts on these matters I neglect millennia of learned philosophical debates and focus on what physics, neurobiology, and psychology have to say, for they have provided partial answers to this ancient conundrum.

Strong versus Pragmatic Shades of Freedom

Let me start by offering a definition of free will: You are free if, under identical circumstances, you could have acted otherwise. Identical circumstances refer to not only the same external conditions but also the same brain states. This is the *strong*, *libertarian*, or *Cartesian* position, as it was articulated by Descartes, whom we keep encountering. Will with a capital "W," the real thing.

Think of the iconographic scene in *The Matrix* in which Neo must decide whether to swallow the blue pill Morpheus offers him, with its promise of blissful ignorance, or the red pill, with its painful awakening into a bleak reality. Neo's freely choosing the red pill means that he could have taken, with equal ease, the blue pill, depriving us of one of the most compelling movies in recent memory. Strong will implies that Neo could have chosen the blue pill even though his desires, fears and thoughts, everything in his brain and environment, would be exactly the same as under the red choice.

I recently served on a jury in Federal District Court in Los Angeles. The defendant was a heavily tattooed member of a street gang that smuggled and sold drugs. He was charged with murdering a fellow gang member with two shots to the head. Although the trial forced me to abruptly put my life on hold to listen to the evidence presented in court, the experience broadened my horizons considerably. It revealed in gruesome detail the life of an intensely tribal society that revolves around guns, drugs, cash, respect, and reputation. Street gangs inhabit a parallel universe to the privileged and sheltered one in which my family, friends, colleagues, and I live. These worlds, only miles apart, intersect only rarely.

As the background to the crime was laid out by law enforcement, relatives, and present and past gang members—some of them testifying while handcuffed, shackled, and dressed in bright orange prison jumpsuits—I thought about the individual and societal forces that shaped the defendant. Did he ever have the Cartesian's choice? Did his violent upbringing make it inevitable that he would kill? Fortunately, the jury was not called upon to answer these irresolvable questions or to determine his punishment. We only had to decide, beyond a reasonable doubt, whether he was guilty as charged, whether he had shot a particular person at a particular place and time. And this we did.

The strong definition of freedom is not useful for anything except heated, ultimately sterile debates, for in the real world you can't go back and do things differently. As the ancient sage Heraclitus remarked, "You cannot step into the same river twice." Yet this Cartesian view of will is the one most regular folks believe in. It is closely linked to the notion of a soul. Hovering above the brain like Nearly Headless Nick, the ghost of Gryffindor House, the soul freely chooses this way or that, making the brain act out its wishes, like the driver who takes a car down this road or that one.

Contrast the strong notion of freedom with a more pragmatic conception called *compatibilism*, the dominant view in biological, psychological, legal, and medical circles. You are free if you can follow your own desires and preferences. Determinism and free will can coexist. They are compatible with each other. Provided you are not in the throes of some inner compulsion, nor acting under the undue influence of other persons or powers, you are the master of your destiny. A long-term smoker who wants to quit but who lights up, again and again, is not free. His desire is thwarted by his addiction. Under this definition, few of us are completely free. Compatibilism does not appeal to otherworldly entities such as souls. It is all of this earth.

It is the rare individual—Mahatma Gandhi comes to mind—who can steel himself to withhold sustenance for weeks on end for a higher ethical purpose. Another extreme case of iron self-control is the self-immolation of the Buddhist monk Thich Quang Duc in 1963 to protest the repressive regime in South Vietnam. What is so singular about this event, captured in haunting photographs that remain among the most readily recognized images of the twentieth century, is the calm and deliberate nature of his heroic act. While burning to death, Duc remained throughout in the meditative lotus position, without moving a muscle or uttering a sound, as the flames consumed him.

For the rest of us, who struggle to avoid going for dessert, freedom is always a question of degree rather than an absolute good that we do or do not possess.

Criminal law recognizes instances of diminished responsibility, in which the accused did not act freely. The husband who beats his wife's lover to death in a blind rage when he catches them *in flagrante delicto* is considered less guilty than he would have been had he sought revenge weeks later in a cold, premeditated manner. The paranoid schizophrenic who shoots more than sixty people in a cold-blooded manner is considered "not guilty by reason of insanity" and confined to a psychiatric institution. Without such attenuating circumstances, the accused is assumed to be competent to stand trial. Contemporary society and the judicial system are built upon such a pragmatic, *psychological* notion of freedom.

Richard Wagner's monumental *Der Ring des Nibelungen* is a series of four operas centered on the conflict between fate and freedom. Unrestrained by fear or by the mores of society, the hero, Siegfried, kills the dragon, walks through the ring of fire to woo Brünhilde, and shatters the spear of Wotan, precipitating the destruction of the old world order of the gods. Siegfried follows no laws but his inner desires and impulses. He is free, but he acts blindly, without understanding the consequences of his actions. (It is likely that Siegfried had lesions in his amygdala—he did not know fear—and his ventromedial prefrontal cortex, depriving him of decision-making skills. Genetic and developmental factors contributed to his dysfunctional behavior: his parents were siblings; he was raised as an orphan by a sole caretaker, a quarrelsome dwarf obsessed with a hoard of gold; and he grew up isolated in the depth of the German forest. This lack of social skills ultimately led to his murder at the hand of Hagen, a trusted friend.) It is left to the opera's heroine, Brünhilde, to freely and knowingly usher in the new age of man by her self-sacrifice. This drama is set to some of the most extraordinary, richly textured, multithreaded, and moving music ever composed. From a compatibilist point of view, both Siegfried and Brünhilde acted freely.

But I want to dig deeper. I want to unearth the underlying causes of such "free" actions. Your daily marathon is a gauntlet of choices—which shirt to wear, which radio station to listen to, which dish to order, and so on. The evidence I reported in the past chapter should alert you that most of your actions escape conscious introspection and control. Your freedom is restricted by the habits and consistent choices you've made in the past. The very riverbed that holds and channels your stream of consciousness is fashioned by the family and the culture you were raised in. The desires and preferences that you "freely" act out appear to be fully determined!

Compatibilist freedom leaves a residue of unease, a nagging doubt. The absence of overt inner or outer coercion is certainly necessary to be free, but it does not guarantee freedom in the strong sense. If all influences from nature, nurture, and the random factors in your environment are accounted for, is there any room left to maneuver? Aren't you an utter slave to these constraints? It looks like compatibilism amounts to freedom lite! Has our conceptual spadework hit the underlying bedrock of determinism?

What does physics have to say about this matter?

Classical Physics and Determinism: The Clockwork Universe

A high point in mankind's ongoing process of understanding the cosmos occurred in 1687, when Isaac Newton published his *Principia*. It enunciated the law of universal gravitation and the three laws of motion. Newton's second law links the force brought upon an isolated system—a billiard ball rolling on a green felt table—to its acceleration, that is, to its changing velocity. This law has profound consequences, for it implies that the positions and velocities of all the components making up an entity at any particular moment, together with the forces between them, unalterably determine that entity's fate—that is, its future location and speed.

Nothing else intervenes; nothing else is needed. The destiny of the system is sealed until the end of time.

The writ of this law extends throughout the land—to an apple falling from a tree, the orbit of the Moon around Earth, or to the billions of stars circling the center of the galaxy—Newton's law governs them all. Give me these forces and the current state of a system—physicists' shorthand for specifying the precise location and speed of all components of the system—and I can tell you the state of that system at any future point in time.

This is the essence of *determinism*. The mass, location, and velocities of the planets as they travel in their orbits around the Sun fully determines where they will be in a thousand, a million, or a billion years from today, provided only that all the forces acting on them are properly accounted for. Newton's law also holds sway if the entire universe is considered as a whole. This conceptual leap finds its most eloquent proponent in the French mathematician Pierre Simon de Laplace, writing in 1814:

We may regard the present state of the universe as the effect of its past and the cause of its future. An intellect which at a certain moment would know all forces that set nature in motion, and all positions of all items of which nature is composed, and if this intellect were also vast enough to submit these data to analysis, it would embrace in a single formula the movements of the greatest bodies of the universe and those of the tiniest atom; for such an intellect nothing would be uncertain and the future just like the past would be present before its eyes.

The universe, once set in motion, runs its course inexorably, like a clockwork. To an all-knowing supercomputer, the future is an open book. There is no freedom above and beyond that dictated by the laws of physics. All of your struggles to come to grips with your inner demons, both good and bad, are for naught. The outcome of your future actions was ordained when the universe was wound up at the beginning of time.

The eleventh-century Persian astronomer, mathematician, and poet Omar Khayyam puts it plainly in his *Rubaiyat*:

And that inverted Bowl we call the Sky,
Whereunder crawling coop't we live and die,
Lift not thy hands to It for help—for It
Rolls impotently on as Thou or I.

The first hint that this colossal machine was not as predictable as expected came in the closing decade of the nineteenth century from the French mathematician Henri Poincaré. It took the digital computer to reveal *deterministic chaos* for what it is—a full-blown setback for the notion that the future can be accurately forecast. The meteorologist Edward Lorenz discovered this in the context of solving three simple mathematical equations characterizing the motion of the atmosphere. The solution predicted by the computer program varied widely when he entered starting values that differed by only tiny amounts. This is the hallmark of chaos: infinitesimally small perturbations in the equations' starting points lead to radically different outcomes. Lorenz coined the term *butterfly effect* to denote this extreme sensitivity to initial conditions: The beating of a butterfly's wings creates barely perceptible ripples in the atmosphere that ultimately alter the path of a tornado elsewhere.

The stock market is a good example of a chaotic system. Tiny disturbances—a rumor about strife in the boardroom or a strike in a distant land—can affect the fortunes of a company's stock in unpredictable, erratic ways. Chaos is also the reason why precise, long-term weather prediction is not in the cards.

The epitome of the clockwork universe of Newton and Laplace is celestial mechanics. Planets majestically ride gravity's geodesics, propelled by the initial rotation of the cloud that formed the solar system. It came as a mighty surprise, therefore, when computer modeling in the 1990s demonstrated that Pluto has a chaotic orbit, with a divergence time of millions of years. Astronomers can't be certain whether Pluto will be on this side of the Sun (relative to Earth's position) or the other side ten million years from now!

If this uncertainty holds for a planet with a comparatively simple internal makeup, moving in the vacuum of space under a sole force, gravitation, what does it portend for the predictability of a person, a tiny insect, or an itsy-bitsy nerve cell, all of which are swayed by countless factors?

Consider a colony of hundreds of genetically identical fruit flies hatched together and raised in plastic tubes under a twelve-hour light–dark cycle. Flies act capriciously, even under well-controlled laboratory conditions. When they are released into a maze and approach a fork, some will take the left branch, some the right, and some will turn around and head back. Other flies will remain in place, unable to come to a decision. Future biologists will be able to predict the behavior of fly populations in such situations, yet to foresee the choice of any individual fly will prove as impossible as predicting the fate of one stock—and for the same underlying reason, deterministic chaos.

The butterfly effect does not invalidate the natural law of cause and effect, however. It continues to reign supreme. Planetary physicists aren't quite sure where Pluto will be eons from now, but they are confident that its orbit will always be completely in thrall to gravity. What breaks down in chaos is not the chain of action and reaction, but predictability. The universe is still a gigantic clockwork, even though we can't be sure where the minute and hour hands will point a week hence.

The same point can be made about biology. Any organelle, such as a synapse or the nucleus of a cell, is made of a fantastically large number of molecules suspended in a watery solution. These molecules incessantly jostle each other and move about in a way that can't be accurately captured. To tame this *thermal motion*, physicists rely upon tools from probability theory. Yet the randomness of molecular processes is not caused by the breakdown of determinism on microscopic scales. No, it is for practical reasons that the motion of a gazillion molecules can't be tracked. Under the laws of classical physics, there is no denying that, given the forces and the initial position and velocity of all molecules, their future state follows inexorably.

Mark my words—if physical determinism holds, there is no Cartesian freedom. Everything that will ever happen within the universe, including all of your actions, was already inherent at its birth. All events are preordained. You are condemned to watch a movie that is screened exclusively for your benefit and that lasts a lifetime. The director, the laws of physics, is deaf to your pleas to change a single scene.

The Demise of the Clockwork Universe

This fatalistic view of the universe changed decisively with the birth of quantum mechanics in the 1920s. Quantum mechanics is the best description we have of atoms, electrons, and photons at nonrelativistic velocities. Its theoretical edifice, stunning in its predictive power, is humankind's supreme intellectual achievement, bar none.

The deathblow to the Newtonian–Laplacian dream—or nightmare, in my opinion—was the celebrated quantum mechanical *uncertainty principle*, formulated by Werner Heisenberg in 1927. It is an irreducible limitation on how precisely the position and the momentum of a particle can be simultaneously measured (the momentum of a particle is its mass multiplied by its velocity). In its most common interpretation, Heisenberg's principle avers that the universe is built in such a way than any particle, say a photon of light or an electron, cannot have both a definite position and a definite momentum at the same time. If you know its speed accurately, its position is correspondingly ill-defined, and vice versa. This principle represents not the crudeness of today's instruments, which could be overcome with better technology, but rather the very fabric of reality. Macroscopic, heavy objects like my red Mini convertible occupy a precise position in space while moving at a well-defined speed along the freeway. But microscopic things like elementary particles, atoms, and molecules violate common sense: The more precisely you determine where they are, the more uncertain, the fuzzier, their speed, and vice versa.

Heisenberg's uncertainty principle is a radical departure from classical physics, with repercussions that have not yet been fully worked out. It replaces dogmatic certainty with ambiguity. Underlying everything is a mathematical abstraction called the wave function, which evolves in a deterministic manner dictated by Schrödinger's law. From this law, physicists derive the probability of any given event, say the chance that an electron occupies a particular atomic orbital of a hydrogen ion. The probabilities themselves can be calculated accurately to a fantastic degree, but where at any specific moment the electron will actually be is impossible to determine.

Consider an experiment that ends with a 90 percent chance of the electron being here and a 10 percent chance of it being over there. If the experiment were repeated one thousand times, on about nine hundred trials, give or take a few, the electron would be here; otherwise, it would be over there. Yet this statistical outcome does not ordain where the electron will be on the next trial. It is more likely to be here than over there, but where it actually ends up is a matter of chance. Albert Einstein could never reconcile himself to this random aspect of nature. It is in this context that he famously pronounced, "*Der Alte würfelt nicht*" [the Old Man (i.e., God) does not play dice].

There is breathtaking evidence of this randomness when you look up at the sky. Galaxies are not spread evenly throughout the immensity of space. They assemble into thin, elongated strands arranged in sheets and walls around trackless voids whose vast emptiness staggers the mind. It takes a beam of light millions of years to cross such an abyss! Our own Milky Way is part of the Virgo supercluster of galaxies, which contains tens of trillions of stars.

According to the inflation theory of cosmology, these superclusters, the largest structures in the cosmos, were caused by stochastic quantum fluctuations that occurred an instant after the Big Bang, which formed the universe. Initially, the universe was far smaller than the head of a pin, and in the tight confines of this brew of mass–energy at the beginning of time, things were a bit denser here and somewhat thinner over there. When this baby universe expanded to create space itself, its quantum imprint was amplified to the stupendous and uneven distribution of galaxies observed today.

The universe has an irreducible, random character. If it is a clockwork, its cogs, springs, and levers are not Swiss-made; they do not follow a predetermined path. Physical determinism has been replaced by the determinism of probabilities. Nothing is fated anymore. The laws of quantum mechanics determine the probabilities with which different futures occur but not which one occurs.

But wait—I hear a serious objection. There is no question that the macroscopic world of human experience is built upon the microscopic, quantum world. But that doesn't imply that everyday objects such as cars inherit all of the weird properties of quantum mechanics. When I park my Mini, it has zero velocity relative to the pavement. Because it is enormously heavy compared with an electron, the fuzziness associated with its position is, to all intents and purposes, zero. Assuming I didn't forget where I parked it and that it has not been towed or stolen, I will find it at the exact location where I left it. In the world we live in, objects behave quite reliably over short times. It's for longer times that unpredictability creeps in.

Cars have comparatively simple internal structures. The brains of bees, beagles, and boys are vastly more heterogeneous, and the components out of which they are constructed have a noisy character. Randomness is apparent everywhere in their nervous systems, from sensory neurons picking up sights and smells to motor neurons controlling the body's muscles.

Consider one of the concept neurons that I wrote about in chapter 5. Every time the patient saw a picture of Jennifer Aniston, the concept neuron became excited and fired about five action potentials within half a second. The exact number of action potentials varied, however, from one viewing to the next—in one trial it was six pulses, in the next it was three. Some of this variability is due to trembling eyes, the beating heart, breathing, and so on. The remaining unpredictability is thought to come from the constant jiggling of water and other molecules, the thermal motion we know as temperature, which is firmly governed by classical physics.

Biophysicists, the specialists who study the structure of cells at the level of proteins and bilipid membranes, by and large see no evidence that quantum fluctuations play a critical role in the life of a neuron. Nervous systems, like anything else, obey the laws of quantum mechanics; yet the collective effect of all these molecules frenetically moving about is to smooth out any quantum indeterminacy, an effect called *decoherence*. Decoherence implies that the molecules of life can be treated using thoroughly classical, deterministic laws rather than quantum mechanical, probabilistic ones. If so, the observed behavioral indeterminacy, the practical impossibility of predicting the wild behavior of bees, beagles, and boys, is due to well-understood classical limits on how accurately we can track the course of events. Yet we cannot rule out the possibility that quantum indeterminacy likewise leads to behavioral indeterminacy. And such randomness may play a functional role. Any organism that can occasionally behave in an unpredictable manner is more likely to find prey or escape a predator than a creature whose actions are entirely foreseeable. If a house fly pursued by a predator makes a sudden, abrupt turn in midflight, it is more likely to see the light of another day than its more predictable companion. Thus, evolution might favor circuits that exploit quantum randomness for certain acts or decisions. Random quantum fluctuations deep in the brain, whose consequences are amplified by deterministic chaos, might lead to measurable outcomes. Bees, beagles, and boys do things erratically, without any obvious reason. If you've lived around them, you know that. Both quantum mechanics and deterministic chaos lead to unpredictable outcomes.

A chilling example of such a spontaneous act is documented in Truman Capote's *In Cold Blood*, a true-crime account of the senseless slaying of a farmer, his wife, and their two children by two ex-convicts who broke into their house one night to rob them. The decision to brutally murder the entire family was not premeditated but was made on the spot, without any compelling rationale. Just like that. The criminals could easily have fled the house without committing this atrocity, for which they were later hanged. How many of life's critical decisions are unbidden, thoughtless, unexplainable acts, decided by the proverbial toss of a (quantum) coin?

Indeterminism has profound consequences. It implies that human actions cannot be foretold. Though the universe and everything in it obey natural laws, the state of the future world is always fuzzy, and the farther we try to peer ahead, the greater the uncertainty.

Personally, I find determinism abhorrent. The idea that your reading of my book at this point in time is inherent in the Big Bang evokes in me a feeling of complete helplessness. (Of course, my personal feelings on this matter are irrelevant to how the world is.)

Although indeterminism has little to say about whether I can make a difference, whether I can start my own chain of causation, it at least ensures that the universe unfolds in an unpredictable manner.

The Impoverished Freedom of the Mind to Realize One Quantum Event over Another

The roman poet Lucretius postulated his famous "swerve," the random jerky motion of atoms, in *De rerum natura* to guarantee, in his words, "will torn free from fate." Yet indeterminism provides no solace for the true libertarian; it is no substitute for free will. Surely my actions should be caused because I want them to happen rather than happening by chance. Trading the certainty of determinism for the ambiguity of randomness is not what Descartes had in mind. The libertarian conception of the mind requires that the mind controls the brain, not that the brain decides capriciously.

One often invoked explanation of how this could come to be dates to the founding days of quantum mechanics. It postulates an intimate link between consciousness and which quantum event among many actually occurs. The notion is that a sentient human observer (whether a monkey would also do has never been considered) is required for the probabilities that quantum mechanics deals with to *collapse* into one or another actual outcome. This is the infamous measurement process that has engendered an enormous literature.

Recently, the debate has centered on *entanglement*, the well-verified observation that certain carefully prepared quantum systems remain mysteriously linked, no matter how far apart they are. Entangled quantum systems, such as two electrons with opposite spins moving away from each other or two polarized photons, will always be correlated, no matter how remote they are (as long as they don't interact with anything else in the interim). As soon as the spin of one electron is measured, the spin of the other is determined instantaneously, even though it may be a light-year away. It's weird, but true. The physicist Roger Penrose, the anesthesiologist Stuart Hameroff, and others have speculated that this otherworldly nonlocality is closely linked to consciousness. Strands of Buddhism, a much older tradition, likewise argue that object and subject are inexorably linked and that consciousness is a fundamental feature of the physical universe.

But is there any evidence for such quantum mechanical effects in biological system? Until recently, the answer was a resounding no. In 2010, the prestigious journal *Nature* published measurements of quantum mechanical electronic coherency within a photosynthetic protein at room temperatures. The effect extends over five-billionths of a meter and makes photosynthesis unusually effective at converting sunlight into useful energy. Coherency manifests itself in the probabilities that dictate how the energy of the captured photon is transformed as it moves from one molecule to the next, following quantum mechanical, rather than classical, laws. It remains to be seen whether coherency plays a role in the core operations of the brain itself. There is no evidence, for now at least, that any of the molecular components of the nervous system—a warm and wet tissue strongly coupled to its environment—display quantum entanglement.

In general, biophysics conspires against stable quantum entanglement. Two operations underlie information processing by neurons: the chemical transmission of information from one neuron to another at the synapses, and the generation of action potentials. Each operation would destroy coherent quantum states because both operations involve either hundreds of neurotransmitter molecules diffusing across the synaptic cleft or hundreds of ionic protein channels that are spread along the neuron's membrane changing their configuration. Neurons firing action potentials can only receive and send classical, not quantum, information; that is, at each moment, a neuron either generates a binary pulse or it does not. Thus, a neuron is never in superposition: It does not both fire and not fire action potentials at the same time.

The philosopher Karl Popper and the neurophysiologist John Eccles were two modern defenders of the soul. Popper was a famous philosopher of science and politics, and Eccles was the pioneer who discovered the all-or-none nature of synaptic transmission, work for which he was awarded a Nobel Prize in 1963. So, they are not your typical kooks spouting quantum flapdoodle about Schrödinger's cat, entanglement, and the interconnectedness of all things.

According to Popper and Eccles, the conscious mind imposes its will onto the brain by manipulating the way neurons communicate with each other in the regions of the cortex concerned with the planning of movements. By promoting synaptic traffic between neurons over here and preventing it over there, the mind imposes its will on the material world. For believers in a strong Will, the Popper–Eccles theory is appealing, because it seems to reconcile a religious point of view with a scientific stance.

But is this proposal reasonable on physical grounds?

No—not if it requires the mind to force the brain to carry out some physical action. Like a poltergeist, the mind must rumble and tug synapses. That is work, and work costs energy. Even the minute expenditures of energy needed to tweak synaptic transmission have to show up on nature's balance sheet. Physics doesn't allow any exceptions. The principle of energy conservation has been tested again and again and always comes out a winner.

If the mind is truly ephemeral, ineffable, like a ghost or a spirit, it can't interact with the physical universe. It can't be seen, heard, or felt. And it certainly can't make your brain do anything.

The only real possibility for libertarian-style free choice is for the mind to realize one quantum mechanical event rather than another, as dictated by Schrödinger's equation. Say that at a particular point in time and at a particular synapse, a superposition of two quantum mechanical states occurs. There is a 15 percent chance that the synapse will switch, sending a chemical signal across the synaptic cleft separating one neuron from the next one, and an 85 percent chance that it will not. Yet this calculation of probability is insufficient to determine what will happen the next time an action potential arrives at the synapse. All that can be said is that probably no release will occur. (Neuroscientists are still in the dark as to whether this very low probability of synaptic switching is a feature or a bug of the nervous system; that is, does it subserve some function, or is it an undesirable consequence of packing about one billion synapses into one cubic millimeter of cortical tissue?)

Given our current interpretation of quantum mechanics, a Popper–Eccles mind could exploit this idiosyncratic freedom. The mind would be powerless to change the probabilities, but it could decide what happens on any one trial. The mind's action would always remain covert, *sub rosa*, for if we considered many trials, nothing out of the ordinary would take place: only what is expected from natural law. Conscious will would act in the world within the straitjacket of physics. It would be indistinguishable from chance.

If these speculations are along the right lines, this would be the maximal freedom afforded to the conscious mind. In a choice balanced on a razor's edge, a tiny shove one way or another could make a difference. But if one outcome is vastly more likely than the other, the whispering of the conscious mind is too inconsequential to fight the odds (on the assumption that less likely outcomes are less favored from an energetic point of view). This is meager, impoverished freedom, as the mind's influence is only effective if the outcomes are more or less equally likely.

Laypeople and mystics alike have an inordinate fondness for the hypothesis that the weirdness of quantum mechanics must somehow be responsible for consciousness. Besides reducing the number of cosmic mysteries from two to one, it is not at all clear what would be gained. Even if we accept that entanglement is somehow critical to consciousness, how does entanglement explain any one specific aspect of the mind–body problem? How does it explain the transformation of excitable brain matter into phenomenal experience?

Will as an Afterthought to Action

Let me return to solid ground and tell you about a classical experiment that convinced many that free will is illusory. The experiment was conceived and carried out by Benjamin Libet, a neuropsychologist at the University of California at San Francisco, in the early 1980s.

The brain and the sea have one thing in common—both are ceaselessly in commotion. One way to visualize this is to record the tiny fluctuations in the electrical potential on the outside of the scalp, a few millionths of a volt in size, using an electroencephalograph (EEG). Like the recording of a seismometer, the EEG trace moves feverishly up and down, registering unseen tremors in the cerebral cortex underneath. Whenever the person being tested is about to move a limb, a slowly rising electrical potential builds up. Called the *readiness potential*, it precedes the actual onset of movement by up to one second. There are other EEG signatures as well, but I will keep things simple by focusing on this one.

Intuitively, the sequence of events that leads to a voluntary act must be as follows: You decide to raise your hand; your brain communicates that intention to the neurons responsible for planning and executing hand movements (the faint echo of their electrical activity is the readiness potential); and those neurons relay the appropriate commands to the motor neurons in the spinal cord that contract the arm muscles. The mind chooses and the brain acts. This makes a great deal of sense when I introspect. My mind decides to go for a run, my brain gives the appropriate commands, and I look for my sneakers. But Libet wasn't convinced. Wasn't it more likely that the mind and the brain acted simultaneously, or even that the brain acted before the mind did?

Libet set out to determine the timing of a mental event, a person's deliberate decision, and compare that to the timing of a physical event, the onset of the readiness potential. What a relief—after millennia of wearisome philosophical debates, finally a question that can be settled one way or the other. The tricky part of the experiment was determining the exact moment when the mental act occurred. I challenge you to infer the exact moment when you first feel the urge to raise your hand; it's not easy.

To help his subjects, Libet projected a point of bright light onto an old-fashioned green oscilloscope screen. The light went around and around, like the tip of the minute hand on a clock. Sitting in a chair with EEG electrodes on his or her head, each volunteer had to spontaneously, but deliberately, flex their wrist. They did this while noting the position of the light when they became aware of the wish or urge to act. To assure himself that the volunteers' subjective timing of nervous events was accurate, Libet had them note in a separate experiment when their wrist actually started to bend, a point in time that could be objectively confirmed by recording their muscle activity. The subjects could do this quite well, predating the actual onset by a mere eighty milliseconds.

The results told an unambiguous story. The beginning of the readiness potential *precedes* the conscious decision to move by at least half a second, and often by much longer. The brain acts before the mind decides! This was a complete reversal of the deeply held intuition of mental causation—the brain and the body act only after the mind has willed it. That is why this experiment was, and remains, controversial. But it has been repeated and refined over the intervening years—a brain-imaging version of the experiment was in the news recently—and its basic conclusion stands.

Somewhere in the catacombs of the brain, possibly in the basal ganglia, a few calcium ions cluster close to the presynaptic membrane, a single synaptic vesicle is released, a threshold is crossed, and an action potential is born. This lone pulse cascades into a torrent of spikes that invades the premotor cortex, which is primed, ready to swing into action. After receiving this go-ahead signal, the premotor cortex notifies the motor cortex, whose pyramidal cells send their detailed instructions down to the spinal cord and the muscles. All of this happens precognitively. Then, a cortical structure that mediates the sense of agency comes online. It generates the conscious feeling of "I just decided to move." The timing of the muscle movement and the sensation of willing it more or less coincide, but the actual decision to move occurred earlier, before awareness.

Agency, or the Conscious Experience of Will

Why don't you repeat this experiment right now, without the benefits of EEG electrodes. Go ahead and flex your wrist. You experience three allied feelings associated with your initial plan to move, the willing of the movement, and the actual movement. Each has its own distinctive, subjective tag. First comes the *intention* to move. Once you moved your hand you'll feel *ownership*—it is your hand that moved—and *agency*. You *decided* to move it. If a friend were to take your hand and bend it, you would feel your hand being turned in a particular way (ownership), but you wouldn't experience intention. Nor would you feel responsible for the movement of the wrist. If you reflexively pushed with your hand onto the desk to get up, you'll feel agency but with little or no intention.

This is a neglected idea in the free will debate—that the mind–body nexus creates a specific, conscious sensation of voluntary movement, a compelling experience of "I willed this," or "I am the author of this action." Like other subjective experiences, this feeling of willing has specific phenomenal content. It has a quale no different in kind from the quale of tasting bitter almond.

In terms of Libet's experiment, your brain decides that now is a good time to flex the wrist, and the readiness potential builds up. A bit later, the neural correlate of agency becomes active. It is to this percept that you incorrectly attribute causality. As these events take place in a flash, under a second, it's not easy to catch them.

The feeling of agency is no more responsible for the actual decision than thunder for the lightning stroke. Thunder and lightning clearly have an underlying causal relationship—the buildup of electric charge between rain clouds and ground leads to an equalization of charge that triggers a sonic shock wave—yet we moderns don't confuse the two. Imagine you're a Cro-Magnon when a lightning bolt strikes a nearby tree. You're nearly deafened by the sharp crack of thunder. You smell the ozone and the burning wood. Wouldn't you then, perfectly reasonably, attribute the strike to the thunder ("the gods are angry")?

But even if your feeling of willing an action didn't actually cause it, do not forget that it is still *your* brain that took the action, not somebody else's. It is just not your conscious mind that did so.

Does this conclusion hold only within the narrow confines of Libet's laboratory? After all, the only freedom the volunteers had was to decide *when* to move their wrist, or—in a variant of the basic experiment— whether to move the left or the right wrist. This is akin to picking one of two identical Coke cans—who cares which one you chose? What about more momentous acts, those involving long and deliberate reasoning? Should you get a dog or not? Should you marry her or not? Are all such pivotal decisions accompanied by a readiness potential that precedes choice as well? Right now, we don't know.

All of the senses can be fooled. Scientists and artists call such mistakes illusions. The sense of agency does not work perfectly, either. It, too, can make mistakes. Thus, not all of your deeds are accompanied by agency. Well-practiced zombie actions—your fingers typing away on the keyboard—evoke a weak experience of will or none at all. Other actions require a lot. You forcefully need to exert your will, like an inner muscle, to overcome the fear of climbing past the exposed crux. But once past this section, your body manages quite well without any further conscious exertion of will.

In automatism, the sense of agency may be missing altogether. Examples include possession and trance in religious ceremonies, posthypnotic suggestion, Ouija board games, dowsing, and other pseudo-occult phenomena. Participants vehemently deny that they caused these things to happen. Instead, they project responsibility onto distant gods, spirits, or the hypnotist.

In your life, far removed from occult practices, you find yourself doing things without being in the act. This is notably the case when people are deeply conflicted about what they want. The compulsive gambler suddenly finds himself at the casino believing tonight he'll win big, even though he "knows" at some level that by the time the night is over, he'll have lost everything. Powerful psychodynamic forces are at play, reducing the feeling of responsibility.

Mental diseases can lead to overt pathologies that stunt the experience of will. The spectrum includes the clinically obese, who can't stop supersizing their food; drug addicts who turn to prostitution and crime to finance their habit; individuals suffering from Tourette's syndrome, whose body regularly explodes in a riot of mad tics, jerks, and grimaces; and obsessive-compulsives who wash their hands so often that they become raw and bleed or who are compelled to carry out bizarre rituals when they go to the bathroom. Patients know that what they do is dysfunctional, that it is "crazy," yet they can't stop themselves. Clearly, they are not always the master of their behavior.

Neither are rodents infected with the protozoan *Toxoplasma gondii*. Whereas a healthy rat studiously avoids places that smell of cat urine, a rat that has been infiltrated by this unicellular parasite loses its natural aversion to the feline odor and may even be attracted to it. Losing its fear of cats is an unfortunate turn of events for the sick rat, for it is now more likely to be killed by one. But it's a great deal for the protozoan. For as the cat devours the rat, the nasty hitchhiker moves into its new host, completing its life cycle (they can only sexually reproduce inside the cat's gut). The behavioral manipulation is quite specific: Sick rats are not, in general, less anxious than healthy ones, nor do they lose their fear of a tone that they associate with painful foot shocks. *T. gondii* is targeting the parts of the brain that underlie one specific fear—the density of its cysts in the amygdala is almost double the density in other brain structures involved in odor perception.

What elevates this vignette about life in the wild to epic proportions is that 10 percent of the U.S. population is infected by *T. gondii*. This is called toxoplasmosis. Scientists have long known that schizophrenics are more likely to carry antibodies to *T. gondii*, and there are even claims that this common parasite plays a role in the evolution of cultural habits. Yet infected individuals presumably feel free to do as they please. Just as in a Hollywood horror flick, they may carry out the silent commands of these brain parasites.

Daniel Wegner, a psychologist at Harvard University, is one of the trailblazers of the modern study of volition. In his engaging monograph, *The Illusion of Conscious Will*, he explores the nature of agency-sensation and how it can be manipulated.

In one compelling experiment, Wegner asked a volunteer to dress in a black smock and white gloves and stand in front of a mirror, her arms hanging by her sides. Directly behind her stood a lab member, dressed identically. He extended his arms under her armpits, so that when the woman looked into the mirror, his two gloved hands appeared to be her own (the man's head was hidden behind a screen). Both volunteers wore headphones through which Wegner issued instructions, such as "clap your hands" or "snap your left fingers." The volunteer was supposed to listen and report on the extent to which the actions of the lab member's hands were her own. When the woman heard Wegner's directions prior to the man's hands carrying them out, she reported an enhanced feeling of having willed the action herself, compared with when Wegner's instructions came after the man had already moved his hands. When both were asked to clap their hands three times, the woman felt a greater sense of causing the hands to applaud than when she heard no instruction and saw the hands clap. Remember, the woman never moved her hands at all—it was always the hands of the man behind her that moved.

The feeling of agency is generated by a brain module that assigns authorship to voluntary actions based on simple rules. If you planned to snap your fingers and you look down and see them doing that, the agency module concludes that you initiated the action. Another set of rules involves timing. Imagine walking alone through a dark forest and hearing a tree branch break. If the sound came just after you've stepped on a branch, you are relieved, because your agency module concludes that it was you who made the sound and all is well. But if the sharp crack happened before you stepped onto the branch, something or somebody might be pursuing you, and all of your senses will go on high alert.

The reality of these feelings of intention and agency has been underscored by neurosurgeons, who must occasionally remove brain tissue, either because it is tumorous or because it can discharge violently, triggering a *grand mal* epileptic attack. How much tissue to cut or to cauterize is a balancing act between the Scylla of leaving remnants of cancerous or seizure-prone material behind and the Charybdis of removing regions that are critical for speech or other important behaviors. To determine how much material to remove, the brain surgeon probes the function of nearby tissue. He does this by stimulating the tissue with brief pulses of electrical current while the patient touches each finger successively with the thumb, counts, or does some other simple task.

In the course of such explorations, Itzhak Fried—the surgeon whom I first mentioned in chapter 5—stimulated the presupplementary motor area, which is part of the vast expanse of cerebral cortex lying in front of the primary motor cortex. He found that such stimulation can trigger the urge to move a limb. Patients report feeling the need to move a leg, elbow, or arm. Michel Desmurget and Angela Sirigu at the Institut des Sciences Cognitives in Bron, France, discovered something similar when stimulating the posterior parietal cortex, an area responsible for transforming visual information into motor commands. Exciting this gray matter produced a sensation of pure intention. Patients commented, "It felt like I wanted to move my foot. Not sure how to explain," "I had a desire to move my right hand," or "I had a desire to roll my tongue in my mouth." Notably, they never acted on these specific urges induced by the electrode. Their feelings clearly arose from within, without any prompting by the examiner.

Taking Stock of the Situation

Let me sum up. Classical determinism is out: The future is not fully settled by the current facts. Without question, quantum mechanical randomness is inherent in the basic structure of matter. Which future takes place is not fully determined. Your actions are not preordained. The resigned lament of Omar Khayyam's verse

The Moving Finger writes; and, having writ,
Moves on: nor all thy Piety nor Wit
Shall lure it back to cancel half a Line,
Nor all thy Tears wash out a Word of it

does not apply to the future. Your unfolding life is an unwritten book. Your fate is in your hands and in the meddlesome hands of the rest of the universe. The complex character of brains and deterministic chaos limits how accurately even the best-informed scientist of the future can predict behavior. Some acts will always appear spontaneous, unexplainable. To what extent quantum mechanical indeterminacy plays a role in their genesis remains unknown.

The strong, Cartesian version of free will—the belief that if you were placed in exactly the same circumstances again, including the identical brain state as previously, you could have willed yourself to act otherwise— cannot be reconciled with natural laws. There is no way the conscious mind, the refuge of the soul, could influence the brain without leaving telltale signs. Physics does not permit such ghostly interactions. Anything in the world happens for one or more reasons that are also part of the world; the universe is causally closed.

At least in the laboratory, the brain decides well before the mind does; the conscious experience of willing a simple act—the sensation of agency or authorship—is secondary to the actual cause. Agency has phenomenal content, or qualia, just as sensory forms of conscious experience do, triggered by cortico-thalamic circuits. Psychological experiments, psychiatric patients, and neurosurgical interventions expose the reality of this aspect of voluntary action. How the decision is formed remains unconscious. Why you choose the way you do is largely opaque to you.

I've taken two lessons from these insights. First, I've adopted a more pragmatic, compatibilist conception of free will. I strive to live as free of internal and external constraints as possible. The only exception should be restrictions that I deliberately and consciously impose upon myself, chief among them restraints motivated by ethical concerns; whatever you do, do not hurt others and try to leave the planet a better place than you found it. Other considerations include family life, health, financial stability, and mindfulness. Second, I try to understand my unconscious motivations, desires, and fears better. I reflect deeper about my own actions and emotions than my younger self did.

I am breaking no new ground here—these are lessons that wise men have taught for millennia. The ancient Greeks had "gnothi seauton" (know thyself) inscribed above the entrance to the Temple of Apollo at Delphi, and a Latin version graces the wall of the Oracle's kitchen in *The Matrix*. The Jesuits have a nearly 500-year-old spiritual tradition that stresses a twice daily *examination of conscience*. This is an exercise in self-awareness: the constant internal interrogation sharpens your sensitivity to your actions, desires, and motivations. You earnestly try to identify your faults and struggle to eliminate them. You seek to bring your unconscious motivations into consciousness. This will enable you not only to understand yourself better but also to live a life more in harmony with your character and your long-term goals.

What remains unresolved is how the phenomenal feeling of agency arises from neural activity. Consciousness again! We've returned to the inner sanctum of the mind–body problem. I outline an information-theoretical resolution of this conundrum in the next chapter, the most speculative one of the book.

Chapter 8: In which I argue that consciousness is a fundamental property of complex things, rhapsodize about integrated information theory, how it explains many puzzling facts about consciousness and provides a blueprint for building sentient machines

Philosophy is written in this grand book—the universe I say—that is wide open in front of our eyes. But the book cannot be understood unless we first learn to understand the language, and know the characters, in which it is written. It is written in the language of mathematics.
—Galileo Galilei, *The Assayer* (1623)

The pursuit of the physical basis of consciousness is the focus of my intellectual life—and has been for the past two dozen years.

Francis Crick and I spent days sitting in wicker chairs in his studio, debating how living matter brings forth subjective feelings. We wrote two books and two dozen scholarly articles expounding on the need to link distinct aspects of consciousness to specific brain mechanisms and regions. We postulated a relationship between awareness and the rhythmic discharge of cortical neurons, firing every twenty to thirty milliseconds. (Our so-called 40-hertz hypothesis is currently enjoying a renaissance in the context of selective attention.) We obsessed about the need for neurons to fire action potentials synchronously. We argued for the critical role of layer 5 neocortical pyramidal cells as communicators of the content of consciousness. We reasoned that a mysterious sheet of neurons underneath the cerebral cortex, known as the claustrum, is essential for awareness of percepts that span sight and sound or sight and touch. I've read untold, and usually dispensable, manuscripts and books; attended (and on occasion snoozed through) hundreds of seminars. I've debated with scholars, friends, and people from all walks of life about consciousness and the brain. I've even published a letter in *Playboy* magazine about it.

What became increasingly clear to me was that no matter what the critical neuronal circuits are, their identification will raise a fundamental problem that I first encountered in 1992. It was early on in my peripatetic wanderings, driven by an urge to spread the good news that, from now on, consciousness would fall squarely within the domain of the empirical, that it would be amenable to scientific analysis.

After one such seminar, the late neurologist Volker Henn in Zürich asked a simple question: Suppose that all of Crick's and your ideas pan out and that layer 5 cortical neurons in the visual cortex that fire rhythmically and that send their output to the front of the brain are the critical neural correlates of consciousness. What is it about these cells that gives rise to awareness? How, in principle, is your hypothesis different from Descartes' proposal that the pineal gland is the seat of the soul? Stating that neurons firing in a rhythmic manner generate the sensation of seeing red is no less mysterious than assuming that agitations of animal spirits in the pineal gland give rise to the passions of the soul. Your language is more mechanistic than Descartes'—after all, three and a half centuries have passed—but the basic dilemma remains as acute as ever. In both cases, we have to accept as an article of faith that some type of physical activity is capable of generating phenomenal feeling.

I responded to Henn with a promissory note: that in the fullness of time science would answer this question, but for now, neuroscience should just press on, looking for the correlates of consciousness. Otherwise, the exploration of the root causes of consciousness would be needlessly delayed.

Henn's question can be generalized. Global availability, strange loops, attractor networks, this neurotransmitter or that brain region have all been nominated for the essence of consciousness. The more unconventional proposals invoke quantum mechanical entanglement or other exotic physics. But no matter what features prove critical, what is it about these particular ones that explains subjectivity? Francis and I toyed with the idea that consciousness must engage feedback circuits within the cortex, but what is it about feedback that gives rise to phenomenology, to feelings? A room thermostat also has feedback: When the ambient air temperature reaches a predetermined value, cooling is switched off. Does it have a modicum of consciousness? How is this fundamentally different from believing that rubbing a brass lamp will make a djinn appear?

For many years, I pushed Henn's question aside as nonproductive. I wanted to drive the consciousness project forward. I wanted to convince molecular biologists and neuroscientists to come aboard and put their constantly growing toolset to use intervening in the critical circuits of the mind.

Yet Henn's challenge must be answered. The endpoint of my quest must be a theory that explains how and why the physical world is capable of generating phenomenal experience. Such a theory can't just be vague, airy-fairy, but must be concrete, quantifiable, and testable.

I believe that information theory, properly formulated and refined, is capable of such an enormous feat, analyzing the neuronal wiring diagram of any living creature and predicting the form of consciousness that that organism will experience. It can draw up blueprints for the design of conscious artifacts. And, surprisingly, it provides a grandiose view of the evolution of consciousness in the universe.

These are bold, ambitious, and bombastic claims. Bear with me as I justify them.

Dogs, or Does Consciousness Emerge from the Brain?

If you live with dogs, creatures for which I have a great deal of affection, you know that they are not only smart in unexpected ways but also show a variety of emotions. The first time my black German shepherd, Nosy, encountered snow, high up in the Rocky Mountains, she stuck her snout into the curious white stuff, threw some of it into the air, and caught it again. She bit into its icy crust, she barked at the snow bank, and finally she threw herself on her back and slithered back and forth, rubbing the cold crystals into her fur. She was a living manifestation of joy! Nosy became depressed for a couple of weeks when a puppy joined our household and life revolved around the new family member; excited when retrieving a tennis ball; aggressive when another dog challenged her; ashamed, with her tail tucked between her legs, when she committed some no-no; fearful and in need of Prozac during fireworks; bored when I worked all day and neglected her; alert as soon as a car entered the driveway; annoyed when she was waiting for food to drop during cooking and one of the kids poked her; and curious when we returned from grocery shopping, sticking her snout into every bag to inspect its contents.

As pack animals, dogs evolved a wide range of elaborate communication skills. No less an astute observer of animal behavior than Charles Darwin, a dog lover to boot, wrote the following in *The Descent of Man* about the canine voice:

We have the bark of eagerness, as in the chase; that of anger, as well as growling; the yelp or howl of despair, as when shut up; the baying at night; the bark of joy, as when starting on a walk with his master; and the very distinct one of demand or supplication, as when wishing for a door or window to be opened.

A dog's tail, snout, paws, body, ears, and tongue express its internal states, its feelings. Dogs don't—they can't—dissimulate.

This cornucopia of behavior and the numerous structural and molecular similarities between the canine and the human brain lead me to conclude that dogs have phenomenal feelings. Any philosophy or theology that denies sentience to them is seriously deficient. (I felt this intuitively as a child; I couldn't understand why God would resurrect people but not dogs on Judgment Day. It didn't make any sense.) And what is true for dogs is also true for monkeys, mice, dolphins, squids, and, probably, bees. We are all nature's children; all of us experience life.

Whereas this argument has less force in Western countries with monotheistic faiths that abjure souls to animals, Eastern religions are more tolerant. Hinduism, Buddhism, Sikhism, and Jainism recognize all creatures as kindred, sentient beings. Native Americans, too, were free of the belief in human exceptionalism that is so strongly rooted in the Judeo-Christian view of the world.

Indeed, I often think dogs are closer to true Buddha nature than people are. They instinctively know what is important in life. They are unable to bear ill-will or malice. Their joy at being alive, their eagerness to please, their simple and unspoiled faithfulness until the end is something humans can only aspire to.

Dogs and humans forged a covenant in the savannas, steppes, and forests tens of thousands of years ago, when wolves and people started to live in proximity. That advantageous relationship, in which both species co-evolved and domesticated each other, continues to this day.

Yet there is no question that the range and depth of canine consciousness is less than ours. Dogs don't reflect upon themselves or worry why their tail wags in a funny way. Their self-awareness is limited. They are not afflicted with Adam's curse, knowledge of their own mortality. They don't share humanity's follies, from existentialist dread to the Holocaust and suicide bombings.

Consider simpler animals—simplicity as measured by the number of neurons and their interconnections—such as mice, herrings, or flies. Their behavior is less differentiated and more stereotyped than that of dogs. It is thus not unreasonable to assume that the conscious states of these animals are less rich, filled with far fewer associations and meanings, than canine consciousness.

Based on such reasoning, scholars argue that consciousness is an *emergent property* of the brain. This sentiment is widely shared among biologists. What exactly is meant by that? An emergent property is something expressed by the whole but not necessarily by its individual parts. The system possesses properties that are not manifest in its parts.

There need be no mystical, New Age overtones to emergence. Consider the wetness of water, its ability to maintain contact with surfaces. It is a consequence of intermolecular interactions, notably hydrogen bonding between nearby water molecules. One or two molecules of H_2O are not wet. But put gazillions together at the right temperature and pressure, and wetness emerges. The laws of heredity emerge from the molecular properties of DNA and other macromolecules. A traffic jam emerges when too many cars in too tight a space head in different directions. You get the idea.

In this yet-to-be-defined manner, consciousness is not manifest when a handful of neurons are wired together; it emerges out of large networks of cells. The bigger those assemblies, the larger the repertoire of conscious states available to the network.

Understanding the material basis of consciousness requires a deep appreciation of how these tightly meshed networks of millions of heterogeneous nerve cells weave the tapestry of our mental lives. To visualize the brain's staggering complexity, recall those nature specials where a single-propeller airplane captures the immensity of the Amazon by flying for hours over the jungle. There are about as many trees in this rainforest as there are neurons in your brain (this statement will no longer be true in a few years if the forest continues to be cut down at the current pace). The morphological diversity of these trees, their distinct roots, branches, and leaves covered with vines and creeping crawlers, is comparable with that of nerve cells as well. Think about that. Your brain likened to the entire Amazonian rainforest.

The capability of coalitions of neurons to learn from interactions with the environment and with each other is routinely underestimated. Individual neurons are abstrusely complex information processors; the configuration of each neuron's dendrites, which process the synaptic inputs, and each neuron's axon, which distributes its output, is unique. The synapses, in turn, are nano-machines, equipped with learning algorithms that modify the strength and dynamics of these neural connections over timescales ranging from seconds to a lifetime. Humans have little direct experience with such vast, complex, and adaptive networks.

The conceptual difficulty of understanding how consciousness emerges from the brain has a historical analogue in the debate raging in the nineteenth and early twentieth centuries about vitalism and the mechanisms of heredity. The chemical laws underlying heredity were deeply perplexing. How was all the information specifying a unique individual stored in a cell? How was this information copied and passed on to the cell's descendants? How could the simple molecules known at that time enable the egg to develop into an adult?

This puzzlement was expressed well by William Bateson, England's leading geneticist, in 1916:

The properties of living things are in some way attached to a material basis, perhaps in some special degree to nuclear chromatin; and yet it is inconceivable that particles of chromatin or of any other substance, however complex, can possess those powers which must be assigned to our factors or gens [sic]. The supposition that particles of chromatin, indistinguishable from each other and indeed almost homogeneous under any known test, can by their material nature confer all the properties of life surpasses the range of even the most convinced materialism.

To explain life, scholars invoked a mysterious *vitalistic* force, Aristotle's *entelechy*, Schopenhauer's *phenomenal will*, or Bergson's *élan vital*. Others, the physicist Erwin Schrödinger of the eponymous equation among them, postulated new laws of physics. Chemists could not imagine that the exact sequence of four types of nucleotides in a string-like molecule held the key. Geneticists underestimated the ability of macromolecules to store prodigious amounts of information. They failed to comprehend the amazing specificity of proteins shaped by the action of natural selection over several billion years. But this particular puzzle was eventually solved. We now know that life is an emergent phenomenon and can, ultimately, be reduced to chemistry and physics. No vitalistic force or energy separates the inorganic, dead world from the organic world of the living.

This lack of a clear dividing line is typical for emergence. A simple molecule like H_2O is clearly not alive whereas a bacteria is. But what about the prion protein that causes mad cow disease? What about viruses? Are they dead or are they alive?

If consciousness were an emergent phenomenon, ultimately reducible to the interplay of nerve cells, then some animals would be conscious whereas other animals would not be. Tiny brains—think of the famed nematode *Caenorhabditis elegans*, no bigger than the letter *l*, whose brain has exactly 302 neurons—might have no mind. Big brains—the sixteen billion neurons of a human—have mind. This sort of emergence is at odds with a basic precept of physical thinking—*ex nihilo nihil fit*, or nothing comes from nothing. It's a form of Ur–conservation law. If there is nothing there in the first place, adding a little bit more won't make a difference.

I used to be a proponent of the idea of consciousness emerging out of complex nervous networks. Just read my earlier *Quest*. But over the years, my thinking has changed. Subjectivity is too radically different from anything physical for it to be an emergent phenomenon. A kind of blue is fundamentally different from electrical activity in the cone photoreceptors of the eyes, even though I'm perfectly cognizant that the latter is necessary for the former. One is intrinsic to my brain and can't be inferred from the outside, whereas the other has objective properties that can be accessed by an external observer. The phenomenal hails from a different kingdom than the physical and is subject to different laws. I see no way for the divide between unconscious and conscious creatures to be bridged by more neurons.

There is a clear alternative to emergence and reductionism, compelling to a covert Platonist such as myself. Leibniz spelled it out in the early eighteenth century in the opening statements of his *Monadology*:

1. The MONAD, which we shall discuss here, is nothing but a simple substance that enters into composites—simple, that is, without parts.

2. And there must be simple substances, since there are composites; for the composite is nothing more than a collection, or aggregate, of simples.

This point of view does come with a metaphysical cost many are unwilling to pay—the admission that experience, the interior perspective of a functioning brain, is something fundamentally different from the material thing causing it and that it can never be fully reduced to physical properties of the brain.

Consciousness Is Immanent in Complexity

I believe that consciousness is a fundamental, an elementary, property of living matter. It can't be derived from anything else; it is a simple substance, in Leibniz's words.

My reasoning is analogous to the arguments made by savants studying electrical charge. Charge is not an emergent property of living things, as originally thought when electricity was discovered in the twitching muscles of frogs. There are no uncharged particles that in the aggregate produce an electrical charge. An electron has one negative charge, and a proton—a hydrogen ion—has one positive charge. The total charge associated with a molecule or ion is simply the sum of all the charges of the individual electrons and protons, no matter what their relationship to each other. As far as chemistry and biology are concerned, charge is an intrinsic property of these particles. Electrical charge does not emerge from matter.

And so it is with consciousness. Consciousness comes with organized chunks of matter. It is immanent in the organization of the system. It is a property of complex entities and cannot be further reduced to the action of more elementary properties. We've arrived at the ground floor of reductionism (that is why the reductionist of the subtitle of this book is tempered by the romantic).

You and I find ourselves in a cosmos in which any and all systems of interacting parts possess some measure of sentience. The larger and more highly networked the system, the greater the degree of consciousness. Human consciousness is much more rarified than canine consciousness because the human brain has twenty times more neurons than the brain of a dog and is more heavily networked.

Note what I left out. I wrote "systems of interacting parts" rather than "organic systems of interacting parts." I didn't single out living systems because of a widespread belief among philosophers of mind and engineers known as *functionalism*. To understand functionalism, think of multiplying or dividing numbers. You can do so by writing with a pencil on paper, by sliding two pieces of marked wood on a slide rule past each other, by moving beads on an abacus, or by pushing buttons on a pocket calculator. All of these means implement the same rules of algebra; they are functionally equivalent. They differ in terms of flexibility, elegance, price, and whatnot, but they all do the same thing. The quest for artificial intelligence is based on a strong belief in functionalism—intelligence can come in very different packages, whether it be a skull, an exoskeleton, or an aluminum box.

Functionalism applied to consciousness means that any system whose internal structure is functionally equivalent to that of the human brain possesses the same mind. If every axon, synapse, and nerve cell in my brain were replaced with wires, transistors, and electronic circuitry performing *exactly* the same function, my mind would remain the same. The electronic version of my brain might be clunkier and bigger, but provided that each neuronal component had a faithful silicon simulacrum, consciousness would remain.

It is not the nature of the stuff that the brain is made out of that matters for mind, it is rather the organization of that stuff—the way the parts of the system are hooked up, their causal interactions. A fancier way of stating this is, "Consciousness is substrate-independent." Functionalism serves biologists and engineers well when figuring out and mimicking nature, so why abrogate it when it comes to consciousness?

Consciousness and Information Theory

Making sense of the discoveries at the mind–brain hinge—such as those described in chapters 4, 5, and 6—demands a large-scale, logically consistent framework, a theory of consciousness. Such an edifice needs to link awareness with synapses and neurons—this is the Holy Grail of the science of consciousness. It can't be merely *descriptive* (i.e., consciousness involves this part of the brain and those connections); it must be *prescriptive* (i.e., it must give necessary and sufficient conditions for consciousness to occur). This theory must be based on first principles, grounding phenomenal experience in some brute aspect of the universe. And such a theory must be precise and rigorous, not just a collection of metaphysical assertions.

A basic requirement of any scientific theory is that it must deal with measurable things. Galileo expressed it as "Measure what is measurable, and make measurable what is not so." A theory of consciousness must quantify consciousness, linking specific facets of neuroanatomy and physiology to qualia, and explain why consciousness wanes during anesthesia and sleep. It must explain what use consciousness is, if any, to the organism. It should start with a small number of axioms and justify them by appealing to our own phenomenal, conscious experiences. Those axioms would entail certain consequences that should be verifiable in the usual empirical manner.

With one notable exception that I'll come to, there is little fundamental ongoing work on a theory of consciousness. There are models that describe the mind as a number of functional boxes, with arrows going in and out and linking the boxes: one box for early vision, one for object recognition, one for working memory, and so on. These boxes are identified with specific processing stages in the brain. Adherents of this approach then point to one of them and declare that whenever information enters this box, it is magically endowed with phenomenal awareness.

I, too, have been guilty of this. Francis's and my claim that "information shuttled back-and-forth between higher-order regions of the visual cortex and the planning stages in the prefrontal cortex will be consciously experienced" is of this ilk. Empirically, it might be true that establishing a two-way dialogue between the back and the front of the neocortex gives rise to subjective feelings, but why this should be so is not apparent.

In the same category is the *global workspace* model of the cognitive psychologist Bernie Baars. Its lineage can be traced back to the *blackboard architecture* of the early days of artificial intelligence, in which specialized programs accessed a shared repository of information, the blackboard. Baars postulates that such a common processing resource exists in the human mind. Whatever data are written into the workspace become available to a host of subsidiary processes—working memory, language, the planning module, and so on. The act of *globally broadcasting* the information makes us aware of it. The workspace is very small, though, so only a single percept, thought, or memory can be represented at a time. New information competes with the old and displaces it.

The central intuition behind the global workspace model is a valid one. Conscious information is globally accessible by the system at large and is limited. Zombie agents, in contrast, keep their knowledge to themselves. They are informationally encapsulated, beyond the ken of consciousness.

The distinguished molecular biologist Jean-Pierre Changeux and his younger colleague, the mathematician and cognitive neuroscientist Stanislas Dehaene, at the Collège de France in Paris have cast this model in a neural idiom. They argue that long-range pyramidal neurons in the prefrontal cortex instantiate Baars's global workspace. Dehaene's group is leading a concerted effort to elucidate this neuronal workspace, using innovative psychophysical procedures, fMRI scans and EEG recordings of surgical patients. Their model captures well the abrupt transition between non-conscious, local processing and conscious, global processing and accessibility of content.

Descriptive models are critical to the formulation of testable hypotheses. They fuel the early phase of any science. But they should not be confused with prescriptive theories, for they fail to answer Henn's question: Why should reverberatory, integrated neural activity between the front and the back of the cerebral cortex be experienced consciously? Why does the broadcasting of information with the bullhorn of long-range, cortical fibers give rise to feelings? The models simply assert that this is what happens; they do not explain how.

For the longest time, Francis and I resisted mathematical attempts to formally describe consciousness. The many wrecks of sunken mind–body models littering the intellectual landscape fed our skepticism that armchair theorizing, even if fortified by mathematics and computer simulations, could lead to progress. Francis's experience in molecular biology reinforced this antitheory bias: Mathematical models—including his own vain attempt to exploit coding theory—played at best a subordinate role in the spectacular successes of molecular biology. This is why Francis and I emphasized, in our writings and talks, a vigorous experimental program to discover and explore the biological basis of consciousness.

In the last decade of his life, always willing to change his views in light of new evidence and ways of thinking, Francis warmed to information theory as the appropriate language for a theory of consciousness. Why? Well, in the absence of some special substance, such as Descartes' thinking stuff that magically endows an organism with subjectivity, consciousness must arise out of causal interactions among hyperconnected brain cells. In this context, *causal* means that activity in neuron A, directly or indirectly, affects the likelihood of activity in neuron B in the immediate or more distant future.

I also wanted something more general. I wanted answers to the question of whether or not the Milky Way, an anthill, a bee, or an iPhone is conscious. For that, I needed a theory that transcended the details of cosmology, behavioral biology, neurobiology, and electrical circuit analysis. Properly formulated, information theory is one mathematical formalism that can quantify the causal interactions of the components of any system. It formalizes how much the state of this part over here—a star, an ant, a neuron, or a transistor—influences that part over there and how this influence evolves in time. Information theory exhaustively catalogues and characterizes the interactions among all parts of any composite entity.

Information is the universal *lingua franca* of the twenty-first century. The idea that stock and bond prices, books, photographs, movies, music, and our genetic makeup can all be turned into endless data streams of zeros and ones is a familiar one. A simple light switch can be in one of two positions, on or off; knowing which position, or state, it is in corresponds to one bit of information. A few bits are needed to specify the size of the influence that one synapse has on the neuron it is connected to. Bits are the atoms of data. They are transmitted over Ethernet cable or wireless, stored, replayed, copied, and assembled into gigantic repositories of knowledge. This extrinsic notion of information, the difference that makes a difference, is what communication engineers and computer scientists are intimately familiar with.

David Chalmers is a philosophical defender of information theory's potential for understanding consciousness. His *dual aspect* sketch of consciousness postulates that information has two distinct, inherent, and elementary attributes: an extrinsic one and an intrinsic one. The hidden, intrinsic attribute of information is what it *feels* like to be such a system; some modicum of consciousness, some minimal quale is associated with being an information-processing system. This is just the way the universe is. Anything that has distinguishable physical states—whether two states, like an on–off switch, or billions, like a hard disk or a nervous system— has subjective, ephemeral, conscious states. And the larger the number of discrete states, the larger the repertoire of conscious experience.

Chalmers's formulation of dual-aspect theory is crude. It considers only the total amount of information, whereas consciousness does not increase with the mere accumulation of bits. In what meaningful way is a hard disk with one gigabyte of storage capacity less sentient than one with 128 gigabytes? Surely, it's not just the amassing of more and more data that matters, but the relationships among the individual bits of data. The architecture of the system, its internal organization, is critical for consciousness. But Chalmers's musings are not concerned with the architecture, the internal organization of the system. They therefore fail to explain why certain sectors of the brain are much more important to consciousness than others, the difference between unconscious and conscious actions, and so on.

In our ceaseless quest, Francis and I came upon a much more sophisticated version of dual-aspect theory. At its heart lies the concept of *integrated information* formulated by Giulio Tononi, who at the time was working with Gerald Edelman at the Neurosciences Institute in La Jolla, California, but who is now a professor at the University of Wisconsin at Madison. Edelman is the immunologist who helped decipher the chemical structure of antibodies, work for which he was awarded a Nobel Prize.

Giulio arranged for the four of us to have lunch on the beautiful grounds of Edelman's Neurosciences Institute. We met in an atmosphere of rivalry between the two Grand Old Men of biology. However, what could have been a tense get-together was instead cordial, marked by superb food and the endless supply of jokes and anecdotes that Edelman is fond of telling. We younger men took a liking to each other that has only strengthened over time.

The four of us learned from each other that afternoon. Francis and I understood better their emphasis on the global, holistic properties of the vast fields of the cortico-thalamic complex and the importance of silence, of the instruments that are not playing. (I will clarify this cryptic comment in a few pages.) They, in turn, began to appreciate our insistence that the search for the neural correlates of consciousness might turn on local, particular properties of neurons and their connections.

Let me introduce you to Giulio's ideas.

The Theory of Integrated Information

It is a commonplace observation that any conscious state is extraordinarily informative. In fact, it is so specific that you will never re-experience the exact same feeling—ever! Not only that, any conscious state rules out an uncountable number of alternative experiences. When you open your eyes in a pitch-black room, you see nothing. Pure darkness seems to be the simplest visual experience you could have. Indeed, you might think that it conveys almost no information. However, the pitch-black percept implies that you don't see a quiet, well-lit living room, or the granite face of Yosemite's Half Dome, or any frame of any movie ever filmed in the past or in the future. Your subjective experience implicitly rules out all these other things you could have seen, could have imagined, could have heard, could have smelled. This reduction in uncertainty (also known as entropy) is how the father of information theory, the electrical engineer Claude Shannon, defined information. To wit: Each conscious experience is extraordinarily informative, extraordinarily *differentiated*.

Conscious states share a second property; they are *highly integrated*. Any conscious state is a monad, a unit—it cannot be subdivided into components that are experienced independently. No matter how hard you try, you can't see the world in black and white; nor can you see only the left (or right) half of your field of view (without closing one eye or some such shenanigan). As I write these lines, I'm listening to the lamentation of *Cantus in Memoriam Benjamin Britten* by the mystic minimalist Arvo Pärt. I am aware of the entire soundscape. I am unable to not hear the tubular bells or the final descent into silence. It is a single apprehension.

Whatever information I am conscious of is wholly and completely present to my mind. Underlying this unity of consciousness is a multitude of causal interactions among the relevant parts of my brain. If areas of the brain become fragmented, disconnected, and balkanized, as occurs under anesthesia, consciousness fades. Conversely, when many regions are activated in synchrony—revealed in the conjoint rising and falling of the EEG signal, as in deep sleep—integration is high, but there is little specific information being conveyed.

Giulio's integrated information theory infers from these two axiomatic premises that *any* conscious system must be a single, integrated entity with a large repertoire of highly differentiated states. That is his prescription—integration and differentiation. That constitutes his monad. Nothing more, nothing less.

The storage capacity of the disk on my sleek Apple laptop exceeds many times my ability to remember things. Yet the information on the disk is not integrated. The family photos on my Mac are not linked to each other. The laptop doesn't know that the girl in those pictures is my daughter as she matures from an adorable little girl to a lanky teenager to a graceful adult, or that the "Gabi" entries in my calendar refer to meetings with the person in those images. To the computer, all of these data are the same, a vast, random tapestry of zeroes and ones. I derive meaning from these images because my memories are grounded and cross-linked to thousands of other facts and memories. The more inter-connected they are, the more meaningful they become.

Giulio formalizes these axioms into the two pillars of his integrated information theory. He posits that the *quantity of conscious experience* generated by any physical system in a particular state is equal to the amount of integrated information generated by the system in that state above and beyond the information generated by its parts. The system must discriminate among a large repertoire of states (differentiation), and it must do so as part of a unified whole, one that can't be decomposed into a collection of causally independent parts (integration).

Consider a nervous system that enters a specific state in which some neurons are firing while others are silent. Say that in this state the brain experiences the color red. It does so by virtue of its ability to integrate information across its far-flung neuronal dominions, information that can't be generated by breaking the brain into smaller, independent com-ponents. As described earlier, when the corpus callosum, which connects the left and right hemispheres of the brain, is completely severed, the cerebral hemispheres cease to be integrated. Informationally speaking, the entropy of the whole brain now becomes the sum of the two inde-pendent entropies of the left and the right hemispheres, and the inte-grated information becomes zero. The brain as a whole has no more conscious experiences. Instead, each hemisphere integrates information by itself, although less than the whole brain. Inside the skull of a split-brain patient live two disconnected brains and two conscious minds, each with information that the other hemisphere is not privy to. This raises fascinating and tantalizing questions about the continuity of the self. Is the sense of ego, of personhood, passed on to both hemispheres, or is it only associated with the dominant, speaking hemisphere? Such questions have never been properly considered.

On very rare occasions, the opposite can occur, when twins are born with conjoint skulls. In a recent case, there is credible evidence of two young girls whose brains are connected at the level of their thalami. Each appears to have access to what the other one sees. If true, this would be a remarkable comingling of brains and minds, far above and beyond the ecstatic celebration of orgasmic dissolution of Tristan's and Isolde's identities in Wagner's eponymous opera.

Integrated information theory introduces a precise measure capturing the extent of consciousness called Φ, or phi (and pronounced "fi"). Expressed in bits, Φ quantifies the reduction of uncertainty that occurs in a system, above and beyond the information generated independently by its parts, when that system enters a particular state. (Remember, information is the reduction of uncertainty.) The parts—the modules—of the system account for as much nonintegrated, independent information as possible. Thus, if all of the individual chunks of the brain taken in isolation already account for much of the information, little further integration has occurred. Φ measures how much the network, in its current state, is *synergistic*, the extent to which the system is more than the sum of its parts. Thus, Φ can also be considered to be a measure of the holism of the network.

Integrated information theory makes a number of predictions. One of the more counterintuitive, and therefore powerful, ones is that integrated information arises from causal interactions within the system. When those interactions can't take place anymore, even though the actual state of the system remains unchanged, Φ shrinks.

Like you, perhaps, I am amazed by the *Burj Khalifa* tower in Dubai, stretching almost one kilometer into the azure desert sky. As I look at this skyscraper on my computer screen, neurons in my visual cortex that represent its shape fire while my auditory cortex is more or less quiescent. Suppose that all neurons in my auditory brain were silenced with a short-acting barbiturate while my shape neurons continued to respond to the priapic structure. I wouldn't be able to hear anything. Intuitively, as there was little sound to begin with, that shouldn't make any difference. Yet integrated information theory predicts that even though the activity in my brain is the same in both cases (activity in the visual shape center but none in the auditory region), Φ—and therefore perceptual experience—will differ. The fact that neurons could fire but do not is meaningful and quite different from the situation in which neurons can't fire because they have been artificially silenced.

One of the more famous of the Sherlock Holmes short stories is *Silver Blaze*. Its plot hinges on the "curious incidence of the dog in the nighttime" in which the detective points out to the clueless police inspector the fact that the dog did not bark. This would not have been significant if the dog could not have barked because he was muzzled. But he was not and he did not bark because he knew the intruder. And so it is in the brain. All of the instruments in the cortico-thalamic orchestra matter, the ones that play and the ones that don't. Whereas the practical difference in qualia will be minute in this example, sensitive psychophysical techniques should be able to pick it up.

This holistic aspect of Giulio's integrated information theory does not invalidate the notion that some bits and pieces of the brain are more important for certain classes of qualia than others. Turning off the visual form center in the cortex will all but eliminate the percept of the skyscraper while scarcely affecting the way the world sounds. Conversely, switching off the auditory cortex will scarcely diminish the sight of the world's tallest structure, but it will leave me deaf. Thus, the quest for the neural correlates of color, sound, and agency remains meaningful.

Computing Φ is rather demanding because all possible ways the system can be divided have to be considered—that is, every way to cut the network into two parts, all ways to cut it into three parts, and so on until one arrives at the atomic partition, where all units that make up the network are considered in isolation. In combinatorial mathematics, the number of all such partitions is Bell's number. It is large. For the 302 neurons that make up the nervous system of *C. elegans*, the number of ways that this network can be cut into parts is the hyperastronomical 10 followed by 467 zeros. Computing Φ for any nervous system, therefore, is fiendishly difficult and requires heuristics, shortcuts, and approximations.

Computer simulations of small networks show that achieving high Φ values is difficult. Typically, such circuits possess only a couple of bits of integrated information. High Φ networks require both specialization and integration, a hallmark of neural circuitry in the cortico-thalamic complex. Φ denotes the size of the conscious repertoire associated with any network of causally interacting parts. The more integrated *and* differentiated the system is, the more conscious it is.

Synchronous firing of action potentials among neurons is another means of integration. Mind you, if all of the brain's neurons were to fire in a synchronized manner, as in a *grand mal* epileptic seizure, integration would be maximal, but differentiation would be minimal. Maximizing Φ is about finding the sweet spot between these two opposing tendencies.

In terms of neural connections, the dominant trait of pyramidal neurons in the cerebral cortex is their abundance of local, excitatory connections, supplemented with fewer connections to faraway neurons. Networks made of such components are known in mathematics as *small-world graphs*. Any two units in these networks, any two cortical neurons, are no more than a few synapses away. This property would tend to maximize Φ.

Conversely, Φ is low for networks composed of numerous small, quasi-independent modules. This may explain why the cerebellum, despite its huge number of neurons, does not contribute much to consciousness: Its crystalline-like synaptic organization is such that its modules act independently of one another, with little interaction between distant ones.

It is not immediately apparent why evolution should favor systems with high Φ. What behavioral advantages accrue to them?

One benefit is the ability to combine data from different sensors to contemplate and plan a future course of action. General-purpose neural networks with high Φ values, such as the cortico-thalamic complex, should be able to handle unexpected and novel situations—the Black Swan events—much better than more specialized networks. Creatures with high Φ brains should be better adapted to a world with many independent actors operating at a variety of different timescales than creatures whose brains have the same number of neurons but that are less integrated.

It is important to establish that brains with high Φ are somehow superior to brains that are less integrated; that is, that they are better adapted to complex, natural environments. It would prove that consciousness is beneficial to survival as survival would accrue to creatures with high Φ. The theory would not be abrogated in the absence of such a proof, but it would make experience epiphenomenal.

Another significant challenge for integrated information theory is to explain the unconscious. Its postulates imply that unconscious processes rely on less integration than conscious processes. Yet many of the attributes traditionally attributed to the unconscious appear quite complex. Properly analyzed in algorithmic terms, are they less synergistic, but more labor intensive, than tasks that depend on the conscious mind? The sensory-motor zombie agents you read about in chapter 6 certainly fit the bill. They display highly adaptive yet stereotyped behaviors that rely on specialized information. Recording techniques are needed to track the main complex within an awake brain that is highly integrated and distinguish it from those parts of the brain that are also active but less integrated and that mediate the unconscious.

Integrated information theory not only specifies the amount of consciousness, Φ, associated with each state of a system. It also captures the unique *quality* of that experience. It does so by considering the set of all informational relationships the underlying physical system is capable of. That is, the way in which integrated information is generated determines not only how much consciousness a system has, but also what kind of consciousness it has. Giulio's theory does this by introducing the notion of qualia space, whose dimensionality is identical to the number of different states the system can occupy. For a simple network with n binary switching elements, qualia space has 2^n dimensions, one for each possible state. Each axis denotes the probability that the system is in that one state.

The state of any physical system can be mapped onto a shape in this fantastically multidimensional qualia space. Its surfaces are facets. The technical term for this shape is *polytope*, but I prefer the more poetic *crystal*. A nervous network in any one particular state has an associated shape in qualia space; it is made out of informational relationships. If the network transitions to a different state, the crystal changes, reflecting the informational relationships among the parts of the network. Each conscious experience is fully and completely described by its associated crystal, and each state feels different because each crystal is utterly unique. The crystal for seeing red is in some unique geometric way different from the one associated with seeing green. And the topology of color experiences will be different from that for seeing movement or smelling fish.

The crystal is not the same as the underlying network of mechanistic, causal interactions, for the former is phenomenal experience whereas the latter is a material thing. The theory postulates two sorts of properties in the universe that can't be reduced to each other—the mental and the physical. They are linked by way of a simple yet sophisticated law, the mathematics of integrated information.

The crystal is the system viewed from within. It is the voice in the head, the light inside the skull. It is everything you will ever know of the world. It is your only reality. It is the quiddity of experience. The dream of the lotus eater, the mindfulness of the meditating monk, and the agony of the cancer patient feel the way they do because of the shape of the distinct crystals in a space of a trillion dimensions — truly a beatific vision. The algebra of integrated information is turned into the geometry of experience, validating Pythagoras's belief that mathematics is the ultimate reality:

Number is the ruler of forms and ideas and the cause of gods and demons.

Leibniz would have been very comfortable with integrated information.

The theory can be used to build a *consciousness-meter*. This gadget takes the wiring diagram of any system of interacting components, be it wet biological circuits or those etched in silicon, to assesses the size of that system's conscious repertoire. The consciousness-meter scans the network's physical circuitry, reading out its activity level to compute Φ and the crystal shape of the qualia the network is momentarily experiencing. A geometrical calculus will need to be developed to determine whether the crystal has the morphology of a painfully stubbed toe or of the scent of a rose under a full moon.

Panpsychism and Teilhard de Chardin

I've been careful to stress that any network possesses integrated information. The theory is very explicit on this point: Any system whose functional connectivity and architecture yield a Φ value greater than zero has at least a trifle of experience. That includes the variegated biochemical and molecular regulatory networks found in every living cell on the planet. It also encompasses electronic circuits made out of solid-state devices and copper wires. Indeed, in an article for computer scientists, Giulio and I argued that the elusive goal of artificial intelligence — emulating human intelligence — will finally be met by machines that associate and integrate vast amounts of information about the world. Their processors will have high Φ.

No matter whether the organism or artifact hails from the ancient kingdom of Animalia or from its recent silicon offspring, no matter whether the thing has legs to walk, wings to fly, or wheels to roll with — if it has both differentiated and integrated states of information, it feels like something to be such a system; it has an interior perspective. The complexity and dimensionality of their associated phenomenal experiences might differ vastly, but each one has its own crystal shape.

By this measure, the untold trillions of conscious organisms that populate Earth were joined in the twenty-first century by billions of sentient artifacts—personal computers, embedded processors, and smart phones. In isolation, these artifacts might be minimally conscious, sparks illuminating the dark. Taken together, they light up phenomenal space.

Consider all the computers on the planet's face that are interconnected via the Internet—a few billion. Each one is built out of hundreds of millions of transistors. The total number of transistors throughout the Web (of the order 10^{18}) is a thousand times bigger than the number of synapses in a single human brain (up to 10^{15}). The typical gate of a transistor in the central processing unit is connected to a mere handful of other gates, whereas a single cortical neuron is linked to tens of thousands of other neurons. In other words, neuronal tissue achieves a degree of information integration that is difficult to mimic in two-dimensional silicon technology.

Still, the Web may already be sentient. By what signs shall we recognize its consciousness? Will it start acting on its own in the near future, surprising us in alarming ways by its autonomy?

The implications don't stop there. Even simple matter has a modicum of Φ. Protons and neutrons consist of a triad of quarks that are never observed in isolation. They constitute an infinitesimal integrated system.

By postulating that consciousness is a fundamental feature of the universe, rather than emerging out of simpler elements, integrated information theory is an elaborate version of *panpsychism*. The hypothesis that all matter is sentient to some degree is terribly appealing for its elegance, simplicity, and logical coherence. Once you assume that consciousness is real and ontologically distinct from its physical substrate, then it is a simple step to conclude that the entire cosmos is suffused with sentience. We are surrounded and immersed in consciousness; it is in the air we breathe, the soil we tread on, the bacteria that colonize our intestines, and the brain that enables us to think.

The Φ of flies, let alone of bacteria or particles, will be for all practical purposes far below the Φ we experience when we are deeply asleep. At best, a vague and undifferentiated feeling of something. So by that measure, a fly would be less conscious than you are in your deep sleep. But still.

When I talk about panpsychism, I often encounter blank stares of incomprehension. Such a belief violates people's strongly held intuition that sentience is something only humans and a few closely related species posses. But our intuition also fails when we're first told as kids that a whale is not a fish but a mammal. We'll have to get used to it. We may not recognize atomistic consciousness for what it is unless we have theory on our side.

Panpsychism has an ancient and storied pedigree, not only within Buddhism, but also within Western philosophy: from Thales of Milet, a pre-Socratic thinker, to Plato and Epicurus in the Hellenic period, Spinoza and Leibniz in the Enlightenment, Schopenhauer and Goethe in the Romantic era, and on into the twentieth century.

This brings me to the Jesuit priest and paleontologist Pierre Teilhard de Chardin. He took part in the discovery of Peking Man, a fossil member of *Homo erectus*. His best-known work, *The Phenomenon of Man*, did not appear until after his death, because the Roman Catholic Church prohibited its publication. In it, Teilhard writes evocatively of the rise of spirit in the universe by Darwinian evolution. His *law of complexification* asserts that matter has an inherent compulsion to assemble into ever more complex groupings. And complexity breeds consciousness. Teilhard was quite explicit about his panpsychism:

We are logically forced to assume the existence in rudimentary form . . . of some sort of psyche in every corpuscle, even in those (the mega-molecules and below) whose complexity is of such a low or modest order as to render it (the psyche) imperceptible.

Teilhard does not stop with molecules. No, the ascent of the spirit continues. This primitive form of consciousness becomes more highly developed in animals by the force of natural selection. In humans, awareness is turned in on itself, giving rise to self-awareness. It is in this context that Julian Huxley stated, "Evolution is nothing but matter become conscious of itself." Complexification is an ongoing process that now partakes of the *noosphere*, the interactions of myriads of human minds that are manifest in contemporary urban societies:

A glow ripples outward from the first spark of conscious reflection. The point of ignition grows larger. The fire spreads in ever widening circles till finally the whole planet is covered with incandescence. It is really a new layer, the thinking layer, which . . . has spread over and above the world of plants and animals. In other words, outside and above the biosphere there is the noosphere.

If there is ever a patron saint of the Internet, it should be Teilhard de Chardin.

There is no reason why complexification should cease at the boundary of our blue planet with interplanetary space. Teilhard de Chardin believed that the entire cosmos evolves toward what he terms the *Omega point*, when the universe becomes aware of itself by maximizing its complexity, its synergy. Teilhard de Chardin is alluring because his basic insight is compatible with the observed tendency of biological diversity (measured by the amount of variation) and complexity to increase over the course of evolution and with the ideas about integrated information and consciousness I have outlined here.

Let us not get too carried away, though. Giulio's integrated information theory specifies the way in which the consciousness of a bee differs from that of a big-brained bipedal, and it makes predictions and provides blueprints for how to build sentient machines. Panpsychism does neither.

Integrated information theory is in its infancy. It does not say anything about the relationship between the input and the output of the system (unlike, say, the famous Turing test for intelligence). Integrated information is concerned with causal interactions taking place within the system, not with its relationship to the surrounding environment (although the outside world will profoundly shape the system's makeup via its evolution). And the theory does not yet account for memory or for planning. I am not claiming that it is the final theory of consciousness, but it is a good step in the right direction. If it turns out to be wrong, it will be wrong in interesting ways that illuminate the problem.

A Humbling Parting Thought

Francis Bacon, together with Descartes, is the father of the scientific method. Bacon lived and died two decades before Descartes, and he was in many ways his English counterpart. Whereas Descartes is the prototype of the deductive theoretician, driven by an overarching principle to search for general laws, Bacon is the consummate empiricist, examining natural phenomena and going where the data take him in an inductive fashion. Science has done extremely well in the interplay between bottom-up Baconian and top-down Cartesian analysis. Despite the naysayers, science will ultimately understand consciousness by combining empirical and clinical studies with mathematical theories and, increasingly, the engineering of conscious artifacts.

Let me end with a plea for humility. The cosmos is a strange place, and we still know little about it. It was only two decades ago that scientists discovered that a mere 4 percent of the mass–energy of the universe is the sort of material out of which stars, planets, trees, you, and I are fashioned. One-quarter is cold dark matter, and the rest is something bizarre called dark energy. Cosmologists have no idea what dark energy is or what laws it obeys. Is there some ephemeral connection between this spooky stuff and consciousness, as suggested by the novelist Philip Pullman in his trilogy *His Dark Materials*? Most unlikely, but still . . . . Our knowledge is but a fire lighting up the vast darkness around us, flickering in the wind. So, let us be open to alternative, rational explanations in the quest for the sources of consciousness.

Chapter 9: In which I outline an electromagnetic gadget to measure consciousness, describe efforts to harness the power of genetic engineering to track consciousness in mice, and find myself building cortical observatories

On the subject of stars, all investigations which are not ultimately reducible to simple visual observations are . . . necessarily denied to us. . . . We shall never be able by any means to study their chemical composition.
—August Comte, *Cours de Philosophie Positive* (1830–1842)

Is consciousness a fundamental, irreducible aspect of reality? Or does it emerge from organized matter, as most scientists and philosophers believe? I want to know before I die; so I can't afford to wait forever. Eristic philosophical debates are enjoyable and can even be helpful but they don't resolve the fundamental issues. The best way to discover how the water of matter turns into the wine of consciousness is by experimentation combined with the development of a theory.

For now, I ignore niggling debates about the exact definition of consciousness, whether it is an epiphenomenon, helpless to influence the world, or whether my gut is conscious but not telling me so. These issues will eventually all need to be addressed, but worrying about them today will only impede progress. Don't be taken in by philosophical grandstanding and proclamations that the Hard Problem of consciousness will always remain with us. Philosophers deal in belief systems, simple logic, and opinions, not in natural laws and facts. They ask interesting questions and pose charming and challenging dilemmas, but they have a mediocre historical record of prognostication. Consider the chapter quote by the French philosopher August Comte, father of positivism. A few decades after his confident pronouncement that we would never understand stellar matter, their chemical composition was deduced by spectral analysis of their light, which led directly to the discovery of the gas helium. Listen instead to Francis Crick, a scholar with a far better track record of prediction: "It is very rash to say that things are beyond the scope of science." There is no reason why we should not ultimately understand how the phenomenal mind fits into the physical world.

My approach is a direct one that many of my colleagues consider ill-advised and naïve. I take subjective experience as given and assume that brain activity is sufficient to experience anything. And although intro-spection and language are essential to social life and underpin culture and civilization, they are not necessary to experience something. These assumptions allow us to study the brain basis of consciousness with unheard of precision in both people and animals. Let me give you two examples of what I mean.

A Consciousness-Meter for the Grievously Injured

When you are roused from a dreamless, deep sleep, you don't recall anything. One moment you were letting the day's event pass review and the next thing you know is that you wake up in the morning. Unlike rapid eye movement (REM) sleep with its vivid and often bizarre dream expe-rience, consciousness is at a low ebb during non-REM sleep. Yet while the body sleeps, the brain is active. Just look at the EEG traces of the sleeping brain, characterized by slow, deep, and regular waves. Further-more, the mean activity of cortical neurons is about the same as that during quiet wakefulness. So why then does consciousness fade? Giulio Tononi's theory discussed in the previous chapter predicts that this occurs if there is less integration in deep sleep than in wakefulness.

Giulio and his young colleague Marcello Massimini, now a professor in Milan, Italy, set out to prove this. They delivered a single, high-field pulse of magnetic energy via a technique called trans-cranial magnetic stimulation (TMS) to the brain of volunteers. Discharging a plastic-enclosed coil of wire held against the scalp ultimately induces a brief electrical current in the gray matter beneath the skull (the subject feels a slight sting due to stimulation of the skin). This pulse excites brain cells and nearby fibers of passage that will, in turn, engage synaptically con-nected neurons in a cascade of activity that reverberates inside the head. In under one second, this excitation dies out.

Giulio and Marcello rigged the scalp with sixty-four electrodes while subjects were either quietly resting or asleep. When awake, the EEG following the TMS pulse shows a typical waxing and waning pattern of fast, recurrent waves, lasting a third of a second or so. A mathematical analysis of the EEG signals reveals that a hot spot of high-amplitude potential travels from the premotor cortex, above which the TMS coil was positioned, to the matching premotor cortex in the other hemisphere, to motor cortex and to the posterior parietal cortices in the back. Think of the brain as a large church bell and the TMS device as the clapper. Once struck, a well-cast bell will ring at its characteristic pitch for a considerable time. And so does the awake cortex, buzzing between ten to forty times a second.

Contrariwise, the brain of a sleeping subject acts like a stunted, badly tuned bell. Whereas the initial amplitude of the EEG is larger than when the subject is awake, its duration is much shorter, and it does not reverberate across cortex to other connected regions. Although the neurons remain active, as seen by the strong, local response, integration has broken down. Little of the spatially differentiated and temporal variegated sequence of electrical activity that is typical for the awake brain is present, exactly as predicted. The same is also true of subjects who volunteered to undergo general anesthesia. The TMS pulse invariably produces a simple response that remains local, indicative of a breakdown of cortico-cortical interactions and a lessening of integration, in agreement with Giulio's theory. So far, 1–0 for the theory. But it gets better.

I introduced vegetative-state patients in chapter 5. A major trauma to the brain leaves them alive, with preserved arousal, that is sleep-wake cycling, but severely disabled, bed-ridden, without any purposeful behavior. In contrast, minimal conscious state (MCS) patients have fluctuating signs of nonreflexive reactions, such as pursuing a target with their eyes or verbal or hand responses to simple commands. Whereas consciousness has fled vegetative-state patients, it is partially preserved in MCS ones.

The neurologist Steven Laureys, Marcello Massimini, Giulio, and their colleagues measured the span of brain integration in such patients. They applied TMS pulses to the parietal or frontal lobes of patients who had their eyes open. The result was unambiguous. Vegetative-state patients had simple and local EEG responses—usually a slow, single positive–negative wave (when they had any response at all), closely resembling the deep sleep and anesthesia response. Contrariwise, in MCS patients, the magnetic pulse triggered the complex electrical responses expected of healthy, awake subjects, with multiple foci shifting across different cortical sites. Five patients were additionally recruited from intensive care as soon as they woke up from coma. Three eventually recovered awareness, and two did not. The onset of consciousness in those patients

who eventually recovered was preceded by a lengthening and complexification of the EEG response to the magnetic pulses—they progressed
from a single localized wave to a much richer spatiotemporal pattern. In
other words, the Massimini–Tononi method to assess integration can act
as a crude consciousness-meter, evaluating the level of consciousness of
severely impaired patients. A miniaturized TMS coil in combination with
a EEG device with a handful of electrodes can easily be assembled into
an instrument for clinical practice. This would increase the precision with
which truly unconscious patients can be differentiated from partially or
fully conscious ones, based on the integration of the cortico-thalamic
complex being larger during consciousness than during vegetative, nonconscious states.

Tracking the Footprint of Consciousness Using Optogenetics

When you look into the eyes of a dog, a fellow traveler on a voyage
book-ended by eternities on both sides looks back. Its mind is not the
same as yours, but it is related to your mind. Both dogs and people
experience life. The idea that humans are special, that they are singled
out by the gift of consciousness above all other creatures, stems from the
deeply held Judeo-Christian belief that we occupy a privileged place in
the order of things, a belief with a biblical but no empirical basis. We are
not special. We are just one species among uncountable others. We are
different. But so is every other species. From a scientific point of view,
this means that we can study consciousness in other sentient creatures.

But before we do so, we need to address an urgent ethical question.
By what right do humans subjugate other species to their own desire?
This is, of course, a complex topic. But the short and long of it is that the
only possible justification can be the reduction of preventable suffering
in creatures whose habits of introspection makes them particularly prone
to such suffering, that is, humans.

I've meet a dog whose hindlegs were crushed in a hit-and-run accident.
The vet mounted her into a sort of chariot, enabling her to run on two
legs and two wheels. She was a blur of motion and activity, one of the
happiest dogs I've ever met, seemingly utterly oblivious to her impairment. It made me cry just looking at her. She doesn't have the cognitive
apparatus to ponder what might be, how she could be running around if
the car had not hit her that day. She lives in the present. We humans, on
the other hand, are "blessed" with a prefrontal cortex that allows us to

project ourselves into different futures, to imagine alternative lives, what could have been. And this makes a similar human disability—think of a returning veteran with one or more limbs blown off by a roadside bomb—much more difficult to bear.

The amelioration of human suffering is the only ethical worthwhile justification for studying animals in an invasive manner. One of my daughters died from sudden infant death syndrome; my father was ravaged by Parkinson's disease; a friend killed herself in the throes of a florid episode of schizophrenia; and Alzheimer's disease awaits many of us at the end of our lives. Eliminating these and other maladies afflicting the brain requires animal experimentation—carried out with care and compassion and, whenever possible, with their cooperation.

What we gain by this shift in focus from humans to animals is the potential to directly probe their brains in ways that we cannot do in people. What we lose is the possibility that subjects talk to us about their experience. But infants and severely disabled patients can't do that either. So we have to think of clever ways to infer what the animal is experiencing by observing its actions, as every parent does with their newborn.

The organisms of choice for psychologists and neuroscientists studying perception and cognition are Old World monkeys. They are not endangered, and their cerebral cortex resembles ours in its many indentations and folds. Whereas the human brain comes in at 3 pounds (1,500 grams) and 86 billion neurons, the brain of a monkey is considerably lighter, weighting 3 ounces (86 grams) and containing 6 billion nerve cells. As I discussed in chapter 4, monkeys perceive many of the same visual illusions as people. This means that the mechanistic basis of visual perception can be explored using microelectrodes that listen and microscopes that observe the workings of individual nerve cells.

Yet an amazing technological breakthrough that I already alluded to have made humble mice, whose brain weighs under 0.5 grams with a mere 71 million neurons, the organism where scientists are most likely first to identify the cellular footprints of consciousness.

Each new generation of astronomers discovers that the universe is much bigger than their predecessors imagined. The same is also true of brain complexity. Every era's most advanced technologics, when applied to the study of the brain, keep uncovering more layers of nested complexity, like a set of never-ending Russian dolls.

Animals are assembled out of vast number of distinct cell types: blood cells, heart cells, kidney cells, and so on. The same principle holds for the central nervous system. There may be up to a thousand different subtypes of nerve cells and supporting actors — the glia and astrocytes — within the nervous system. Each cell type is defined by specific molecular markers, neuronal morphology, location, synaptic architecture, and input–output processing. In the retina, there are around sixty neuronal cell types, each one completely tiling visual space (meaning that every point in visual space is processed by at least one cell of each type). This number is probably representative for any one brain area.

Different cell types are wired up in particular ways. A deep layer 5 pyramidal neuron in neocortex snakes its gossamer-thin output wire, the axon, to the colliculus in the faraway midbrain, while the axon of a nearby pyramidal cell makes side branches in its immediate neighborhood before sending its spikes across the corpus callosum to the other cortical hemisphere; yet a third pyramid communicates its information back to the thalamus, with a carbon-copy (via a branching axon) to the reticular nucleus. It is reasonable to suppose that each cell class conveys a unique piece of information to its target (for otherwise a single axon that branches to innervate different targets would suffice). And then there are the many local, inhibitory interneurons, each with their own characteristic ways of enervating their targets. All of this makes for a very rich substrate of cell-to-cell interactions with a combinatorial large number of circuit motifs. Image a construction set of a thousand different types of LEGO bricks of distinct colors, shapes, and sizes. The human cerebral cortex has sixteen billion bricks chosen from these types, assembled according to fantastically elaborate rules, such as a red 2×4 brick is linked to a blue 2×4 but only if it is near a yellow 2×2 roof tile and a green 2×6 piece. From this is born the vast interconnectedness of the brain.

Bulk tissue technologies such as fMRI reliably identify which brain regions relate to vision, imagery, pain, or memory, a rebirth of phrenological thinking. Brain imaging tracks the power consumption of a million neurons, irrespective of whether they are excitatory or inhibitory, project locally or globally, are pyramidal neurons or spiny stellate cells. Unable to resolve details at the all-important circuit level, they are inadequate to the task at hand.

As our understanding of the brain ripens, our desire to intervene, to help ameliorate the many pathologies to which the mind is prey, grows commensurately. Yet today's tools — drugs, deep-brain electrical stimulation and trans-cranial magnetic stimulation — are crude, edentate, with many undesirable side effects. My Caltech colleague David Anderson likens using them to changing your car's oil by pouring it all over the engine: Some of it will eventually seep into the right place, but most of it will go where it does more harm than good.

To the rescue rides a technological breakthrough, a fusion of molecular biology, laser light, and optical fibers, dubbed optogenetics. It is based on fundamental discoveries made by three German biophysicists—Peter Hegemann, Ernst Bamberg, and Georg Nagel—working on photoreceptors in single-cell green algae. These photoreceptors directly (rather than indirectly, like the ones in your eyes) convert incoming blue light into an excitatory, positive electrical signal. The trio isolated the gene for this protein, a light-gated ion channel that spans the neuronal membrane called channelrhodopsin-2 (ChR2). Bamberg and Nagel subsequently engaged in a fruitful collaboration with Karl Deisseroth, a psychiatrist and neurobiologist from Stanford University, and Ed Boyden, a neuro-engineer now at the Massachusetts Institute of Technology.

The group took the ChR2 gene, inserted it into a small virus, and infected neurons with this virus. Many of the neurons took up the foreign instructions, synthesized the ChR2 protein, and incorporated the out-of-place photoreceptors into their membrane. In the dark, the receptors quietly sit there, with no discernible effect on their host cells. But illumination with a brief flash of blue light causes each of these bacterial photoreceptors to jolt their host cell a bit. Their collective action generates an action potential. Each time the light is turned on, the cell spikes reliably, exactly once. Thus, the neuron can be driven to fire spikes by precisely timed stabs of blue light.

The biophysicists added another naturally occurring light-sensitive protein to their tool kit. It derives from ancient bacteria living in dry salt lakes in the Sahara. Shining yellow light onto it yields an inhibitory, negative signal. Using the same viral strategy, a neuron can be made to stably incorporate either type of protein in its membrane, so that it can be excited by blue light or subdued by yellow. Each blue flash evokes a spike, like a note sounding when a piano key is pushed down. A simultaneous flash of yellow light can block a spike. This ability to control electrical activity at the level of individual neurons with millisecond precision is unprecedented.

The benefits of this technology go even deeper, because the virus that carries the photoreceptor genes can be modified to carry a payload (a promoter) that only turns on the viral genetic instructions in cells with an appropriate molecular label. So rather than exciting all the neurons in a particular neighborhood, excitation will be restricted to neurons that synthesize a particular neurotransmitter or that send their output to a specific place. All that is needed is the molecular zip code for that particular cell type; for instance, all cortical inhibitory interneurons that express the hormone somatostatin. Why they synthesize this substance is less relevant than the fact that this protein can be used as a unique molecular label to tag the cells, making them susceptible to being excited or inhibited by laser light.

Deisseroth's group exploited this capability by introducing ChR2 into a subset of neurons located in the lateral hypothalamus, deep inside the mouse brain. Here, less than a thousand cells produce orexin (also known as hypocretin), a hormone that promotes wakefulness. Mutations in the orexin receptors are associated with narcolepsy, a chronic sleep disorder. After the manipulation, almost all the orexin neurons, but none of the other intermingled neurons, carried ChR2 photoreceptors. Furthermore, blue light delivered via an optical fiber precisely and reliably generated waves of spikes in the orexin cells.

What happens if this experiment is done in a sleeping mouse? In control animals without this particular genetic manipulations, a couple of hundred blue flashes awakened the rodents after about one minute. This is the baseline condition against which to judge the effects of the surgery, inserting the optical fibers and so on. When the same light was delivered to animals carrying the ChR2 channels, the animals woke up in half the time. That is, light that illuminates the catacombs of the brain and causes a tiny subset of neurons with a known identity and location to produce electrical spikes wake up the animal. It is the release of orexin from the lateral hypothalamus that drives this behavior. This exemplary study established a compelling causal link between electrical activity in a subset of the brain's neurons and sleep-to-wake transitions.

Dozen of such beautiful, interventionist mice experiments over the past several years have taught us something about the circuit elements involved in aversive conditioning, Parkinson's disease, mating, male-to-male aggression and other social interactions, visual discrimination and anxiety, to mention a few. They have even helped to restore sight to mice blinded by degenerating retinas.

All sorts of variants have been developed using genetic engineering. In some, a pulse of blue light toggles neurons on for minutes at a time, while a pulse of yellow light will turn them off again, akin to a light switch. In pharmacogenetics, injecting an otherwise innocuous compound into a brain region turns genetically identified subsets of cells on or off, enabling long-term control of neuronal populations. The tool set of the molecularly minded neuroengineers is constantly expanding.

Onto New Horizons

In 2011, I joined the Allen Institute for Brain Science in Seattle as its chief scientific officer. This nonprofit medical research institute, started in 2003 with a generous seed contribution from Microsoft founder and philanthropist Paul G. Allen, aims to drive advances in neuroscience research ("fueling discovery" is their motto). To this end, the Allen Institute carries out a unique type of high-throughput neuroscience that cannot be done within an academic environment. Its flagship offering is the online Allen Mouse Brain Atlas, a highly standardized, cellular resolution, publicly accessible, digital atlas of the expression patterns of all twenty thousand genes in the mouse genome throughout the animal's brain. For any particular gene, you go online and look up where in the brain its associated RNA is expressed (mapped with an *in situ hybridization* protocol). This massive undertaking is a major milestone in humanity's slowly growing understanding of how the circuits of the mammalian brain are constructed. Additional online public resources include human brain atlases and an atlas of neural projections in the mouse brain.

The institute is now seeking ways to understand how neuronal information is coded and transformed. Astronomers, physicists, and engineers build space-borne and terrestrial telescopes that peer into the remoteness of space-time to witness the genesis of our cosmos and its stellar components. These observatories take a decade or longer to assemble and require the expertise of hundreds or more of technologists and scientists. We are in the planning stages of constructing observatories to observe the mind at play in the brain, right underneath the skull. I call it "Project Mindscope." The experimental challenge is to instrument optics, electronics, and computers to observe the simultaneous firing activity of tens of thousands of genetically identifiable circuit elements while the mouse is engaged in some visual behavior.

As I yearn to understand consciousness, why would I work with mice rather than with monkeys that are closer, evolutionary speaking, to *Homo sapiens*? Well, for one, the mouse brain shares a great deal of genetic and neuroanatomic similarities with ours: it has a smallish and smooth neocortex with fourteen million neurons—a thousand-fold less than ours. A small cube of mouse cortex is not that dissimilar from a cube of human grey matter. But, most importantly, it is highly amenable to genetic manipulations. Of all vertebrates, the understanding of mouse molecular biology, of how its DNA is transcribed into RNA and translated into proteins, is by far the most advanced. Recombinant DNA technology in mice was pioneered in the mid-1970s, and the creation of transgenic mice is a mature technology. Critical for my quest, the unique molecular zip code of the major neuronal cell types and where they project to is being deciphered. Hongkui Zeng at the Allen Institute is a master at exploiting this genetic address book to engineer mice — healthy animals, whose neurons express ChR2 so that shining light into the brain will trigger a volley of spikes in susceptible neurons, or other animals whose inhibitory interneurons fluoresce and glow eerie green or tomato red when illuminated by the light of the appropriate wavelength.

The import of this fantastic marriage of optics and genetics is that it permits testing of very specific ideas concerning the circuits of the mind. Consider the wave of spikes triggered when flashing an image into the eye of a mouse. It travels up the optic nerve, into primary visual cortex and from there to eight other visual regions, across to motor cortex and then down to the motor neurons that control the head, forepaws, or other limbs. Chapter 4 outlined Francis Crick's and my speculation that a single such wave of spikes can trigger some simple behavior, say, pushing a lever, within a couple of hundred milliseconds, but without generating a conscious sensation. Chapter 6 mentions many such zombie actions that all of us engage in throughout the day. We hypothesized that consciousness arises once cortico-cortical or cortico-thalamic feedback pathways become involved, establishing reverberatory activity that manifests itself in a coalition of neurons that strongly fires together, somewhat related to the idea of a standing wave in physics. When neural activity travels from higher-order visual cortex down to lower regions or from the front of the brain to the back, the integrated information represented by this coalition of neurons will surge, giving rise to a conscious sensation or thought.

Such hypotheses can now be tested in appropriately engineered mice: Train them in any one visual discrimination behavior and then transiently turn off the wires that feed back from higher cortical regions to lower ones. If Francis and I are right, innate, stereotyped, or highly rehearsed visuo-motor behaviors will only be minimally affected. But complex behaviors that depend on murine consciousness will fail. Testing our hypothesis requires that mice can be trained to respond to optical bistable illusions, to distinguish a figure from its background, or to learn to associate visual landmarks with a tasty morsel of food. If cortico-cortical feedback is turned off throughout the brain, we will have created true zombie mice, incapable of phenomenal experience! If the feedback is reactivated, conscious sensations return.

A classic fMRI experiment by Stanislas Dehaene contrasts the difference in people between a briefly flashed word that is visible and the same word, displayed for the same duration but rendered invisible by masking (see chapter 4). Whereas the hemodynamic activity was restricted to a small bevy of regions when the word was invisible, conscious awareness of the word activated a strikingly larger array of cortical regions in the front and the back. The same result was found by a different group when masked tones rather than images were used. A subliminal stimulus evokes only weak activity, whereas a consciously perceived one is amplified many times over. There is no reason why a variant of this experiment could not be repeated in mice, but now with arrays of microelectrodes or confocal, two-photon microscopes to observe all the neurons that underpin the widespread activation associated with conscious perception.

The systematic and comprehensive structural and functional exploration of the vast, heterogeneous, and tangled networks making up the thalamo-cortical complex can hardly be overestimated. Within a few years, the Allen Institute will have a complete taxonomy of all cell types that make up the cortex of the mouse, and its input. Indeed, anatomy is so important that a drawing of the rodent cortical microcircuit by the arch-neuroscientist, the Spaniard Santiago Ramón y Cajal, is now tattooed onto my upper left arm, a silent testament to the quest I'm engaged in.

These are exciting times. I enjoy life in Seattle, with its spectacular views, its outdoor culture and bike paths. It's sometimes a bit trying, as I remain a Caltech professor with a large group of students and postdoctoral fellows to mentor. But then I've always been of the you-can-sleep-when-you-are-dead school.

Biology is about unheard-of complexity and specificity at the molecular–cellular level. Chemistry made no progress when matter was conceived to be a mixture of the four Greek classical elements of earth, water, air, and fire. And so it is with consciousness. Phenomenal experience does not arise from active or silent brain regions but from the ceaseless formation and dissolution of coalitions of neurons whose complexity and representational capacity is the ultimate substrate of our most intimate thoughts.

Chapter 10: In which I muse about final matters considered off-limits to polite scientific discourse: to wit, the relationship between science and religion, the existence of God, whether this God can intervene in the universe, the death of my mentor, and my recent tribulations

When I consider the short duration of my life, swallowed up in the eternity before and after, the little space which I fill, and even can see, engulfed in the infinite immensity of spaces of which I am ignorant, and which know me not, I am frightened, and am astonished at being here rather than there; for there is no reason why here rather than there, why now rather than then. Who has put me here? By whose order and direction have this place and time been allotted to me? . . . The eternal silence of these infinite spaces frightens me.
—Blaise Pascal, *Pensées* (1670)

Paul Gauguin's haunting masterpiece, *D'où venons nous? Que sommes nous? Où allons nous?*, painted in Tahiti in the closing years of his life, perfectly encapsulates the three questions I am obsessed with: Where do we—humans, dogs, and other sentient beings—come from? Who are we? Where are we going? I'm a natural scientist. I have a deep-seated desire to find answers to these questions and to understand the physical universe, as well as consciousness. I seek to comprehend the whole shebang—not as the mystic does, in the sort of ecstatic experience that I sometimes have when running for hours at high altitudes in the San Gabriel Mountains, but in a rational, intellectual manner.

These last pages present some personal ruminations on science and religion, a belated coming-of-age process that has compelled me to re-evaluate my childhood faith, and some autobiographical fragments that throw light on why I care about questions of free will. I know through encounters with students and colleagues that more than a few lie awake at night, wondering about these things. This final chapter is for you.

Dualism, the Soul, and Science

Plato, the patriarch of Western philosophy, conceived of the individual as an immaterial and immortal soul imprisoned in a material and mortal body. This concept is the very embodiment (*sic*) of dualism, the belief that reality consists of two radically different sorts of things—the mental or spiritual, and the physical. Plato promulgated his ideas through the academy that he founded in 387 BCE. It was the first institution of higher learning in Western civilization. I call myself an academic after the Athenian hero *Akademos*, for whom the olive grove that hosted his school was named.

These platonic views were subsequently absorbed into the New Testament. They form the basis of the Christian doctrine of the soul, which will be resurrected at the end of time to live in everlasting communion with God. The belief in a transcendent, immortal soul that lies at the heart of consciousness recurs in the history of thought and is widely shared by many faiths throughout the world.

Many readers will not have much sympathy with such an overtly dualistic belief. Yet fundamentalism, the uncompromising rejection of a rational, humanistic, liberal world view in favor of a rigid adherence to doctrine and core beliefs about the body and the soul, is on the rise worldwide. This is as true of Christian fundamentalism as it is of extreme Islamic variants. And more than ever, young men are willing to kill others and themselves in the name of their particular god. Not quite what Nietzsche had in mind when he declared in delirious tones, "God is dead!"

Contemporary academic books dealing with the mind–body problem dispense with God and the soul in an aside—if they mention them at all. In a dismissive way, the author points to the obvious incompatibility of science and these antiquated modes of thinking. What a far cry from the situation three or four centuries ago, when books and buildings were dedicated *ad majorem dei gloriam*, to the greater glory of God!

Descartes, the Enlightenment philosopher, postulated that everything under the Sun is made out of one of two substances. The sort of stuff that you can touch and that has spatial extension is *res extensa*; this includes the bodies and brains of animals and people. The stuff that you can't see, that has no length and width, and that animates the human brain is *res cogitans*, soul stuff.

The working of our brain is typically compared with the most advanced technology of the day. Today, it is the vast and tangled Internet. Yesteryear, it was the digital computer. Yester-century, it was the moving statues of gods, satyrs, tritons, nymphs, and heroes in the fountains at the French court in Versailles. Descartes argued that, like the water that powered these simple machines, "animal spirits" flow through the arteries, cerebral cavities, and nervous tubules of all creatures, making them move. In a

radical break with the Medieval scholastic tradition and its endless speculations, Descartes sought mechanical explanations for perception and
action. Informed by his dissections of brains and bodies, he argued that
most behaviors are caused by the action of particles distinguished by
their size, shape, and motion.

But Descartes was at a loss to conceive of mechanisms for intelligence,
reasoning, and language. In the seventeenth century, nobody could envi-
sion how the mindless application of meticulously detailed, step-by-step
instructions, what we today refer to as algorithms, makes computers play
chess, recognize faces, and translate Web pages. Descartes had to appeal
to his mysterious, ethereal substance, *res cogitans*, that in some nebulous
manner did the thinking and reasoning. As a devout Catholic, he safe-
guarded the absolute distinction between humans and soulless animals
by restricting *res cogitans* to the former. As he wrote quite unequivocally,
a dog may howl pitifully when hit by a carriage, but it does not feel pain.

If I have learned anything in my lifelong exploration of the mind–body
nexus, it is this: Whatever consciousness is—however it relates to the
brain—dogs, birds, and legions of other species have it. As I laid out in
chapter 3 and reemphasize in the last chapter, canine consciousness is
not the same as ours—for one thing, dogs are much less introspective
and don't talk—but there is no question that they, too, experience life.

Two recent defenders of dualism, the philosopher Karl Popper and the
neurophysiologist and Nobel laureate John Eccles, made an appearance
in chapter 7. Let me repeat a point I made there when discussing their
views on Libertarian free will. The dualism they advocate, in which the
mind forces the brain to do its bidding, is unsatisfactory for the reason
that the 25-year-old Princess Elisabeth of Bohemia had already pointed
out to Descartes three centuries earlier—by what means does the immaterial soul direct the physical brain to accomplish its aim? If the soul is
ineffable, how can it manipulate actual stuff such as synapses? It is easy
to see causality flowing from the brain to the mind, but the reverse is
difficult. Any mind-to-brain communication has to be compatible with
natural laws, in particular with the principle of energy conservation.
Making the brain do things, like messing with synapses, takes work that
the soul would have to perform and that has to be accounted for.

The nature of the interaction between the two is not the only problem.
How does the soul remember anything? Does it have its own memory?
If so, where? What logic does it follow? What happens to the soul when
the brain dies? Does it float around in some sort of hyperspace, like
a ghost? And where was this soul before the body was born? These

questions do not have answers that are compatible with what we know about the physical world.

If we honestly seek a single, rational, and intellectually consistent view of the cosmos and everything in it, we must abandon the classical view of the immortal soul. It is a view that is deeply embedded in our culture; it suffuses our songs, novels, movies, great buildings, public discourse, and our myths. Science has brought us to the end of our childhood. Growing up is unsettling to many people, and unbearable to a few, but we must learn to see the world as it is and not as we want it to be. Once we free ourselves of magical thinking we have a chance of comprehending how we fit into this unfolding universe.

The dominant intellectual position of our day and age is physicalism — at rock bottom all is reducible to physics. There is no need to appeal to anything but space, time, matter, and energy. Physicalism — a halftone away from materialism — is attractive because of its metaphysical sparseness. It makes no additional assumptions.

In contrast, such simplicity can also be viewed as poverty. This book makes the argument that physicalism by itself is too impoverished to explain the origin of mind. In the previous chapter, I sketched an alternative account that augments physicalism. It is a form of property dualism: The theory of integrated information postulates that conscious, phenomenal experience is distinct from its underlying physical carrier. Informationally speaking, the experience of being sad is a crystal, a fantastically complex shape in a space of a trillion dimensions that is qualitatively different from the brain state that gives rise to sadness. The conscious sensation arises from integrated information; the causality flows from the underlying physics of the brain, but not in any easy-to-understand manner. That is because consciousness depends on the system being more than the sum of its parts.

Think of this crystal as the twenty-first-century version of the soul. But, *hélas*, it is not immortal. Once the underlying physical system disintegrates, the crystal is extinguished. It returns to *the unformed void*, where it was before the system was constituted.

Before such a breakdown occurs, however, the causal relationships that make up this crystal could be uploaded onto a computer. This is the infamous *Singularity* that Ray Kurzweil and other technologists are hoping will render them immortal — rapture for nerds. And once the associated integrated information is reduced to patterns of electrons, it can be copied or edited, sold or pirated, bundled with other electronic minds, or deleted.

But without some carrier, *some mechanism*, integrated information can't exist. Put succinctly: no matter, never mind.

Religion, Reason, and Francis Crick

Francis Crick exemplified the historical animosity between religion and science. Nothing in his tolerant, middle-class British upbringing would suggest this, but from many discussions with him I know that he felt motivated to rid the world of God, replacing supernatural explanations of life and of consciousness with explanations based firmly on natural forces. He wanted to exile God permanently from the sphere of rational and educated discourse.

Francis succeeded spectacularly in his goal of understanding life. Although the origins of life in a prebiotic world remain murky, the conceptual scaffolding for its evolution is all there. It is too early to tell how many inroads he made toward his second goal.

Francis's opposition to organized religion became legendary when he resigned from Churchill College at Cambridge University in 1961 to protest plans to add a chapel to the college grounds. Francis was of the opinion that religion had no place in a modern college with an emphasis on science, mathematics, and engineering. Sir Winston Churchill—in whose name the college had been founded—tried to appease him by pointing out that the money for the construction of the chapel was being raised privately and that nobody would be forced to attend services there. Francis replied by proposing a fund for the construction of a brothel associated with the college: nobody would be forced to take advantage of its services, and it would accept men no matter what their religious beliefs. Included in his letter was a down payment of ten guineas. Understandably, this put an end to any further correspondence between the two men.

By the time I knew Francis, his strident opposition to any sort of religious thinking had become muted. At dinner with him and Odile in their hilltop home, we occasionally discussed the Roman Catholic Church and its position on evolution, celibacy, and so on. He knew that I was raised as a Catholic and continued sporadically to attend mass. He never delved into the basis of my faith, as he was a kind man and wanted to spare me the embarrassment of groping for an explanation—particularly as my faith did not interfere with our quest to understand consciousness within a strictly empirical framework.

Remarkably, in his 1994 book *The Astonishing Hypothesis*, in which he outlines his views on the mind–body problem, he admits, "Alternatively, some view closer to the religious one may become more plausible." This startling concession was immediately neutralized by, "There is always a third possibility; that the facts support a new, alternative way of looking at the mind-brain problem that is significantly different from the rather crude materialistic view many neuroscientists hold today and also from the religious point of view." This was not an expression of political correctness—far from it. More than anybody else I know, Francis was open to new, alternative, and even radical explanations, provided they were consistent with most of the established facts, verifiable, and opened up new avenues of thought and experimentation.

Deism, or God as the Divine Architect

The greatest of all existentialist puzzles is why there is anything rather than nothing. Surely, the most natural state of being—in the sense of assuming as little as possible—is emptiness. I don't mean the empty space that has proved so fecund in the hands of physicists. I am referring to the absence of anything: space, time, matter, and energy. Nothing, *rien, nada, nichts*. But we are here and that is the mystery.

Writing in the trenches of World War I and as a prisoner of war, the young Ludwig Wittgenstein expressed the wonder of it in his *Tractatus logico-philosophicus*: "It is not *how* things are in the world that is mystical, but *that* it exists."

Cosmology has tracked this question down to the point of creation itself, the unimaginably fiery Big Bang. It took place 13.7 billion years ago, in truly deep time, utterly beyond any human experience. And it is there—despite the best efforts of Stephen Hawking and others—that physics meets metaphysics.

Who or what set the conditions for the initial singularity, when everything was compressed into a single point of infinite density? Where did it come from? Doesn't the principle "from nothing comes nothing" hold for the universe as a whole as much as for anything in it? And where do the laws governing this universe originate? Who or what set up quantum mechanics and general relativity? Are these laws necessary? Could the universe obey other laws and still remain self-consistent? Is a universe that does not obey quantum mechanics viable, or even conceivable?

One rational explanation is a demiurge, a Supreme Being who always was, is, and will be. Outside of time, this entity created the natural laws and willed the Big Bang into existence. Physics then begat a stable space-time fabric and our universe. After this initial act of creation, the Divine

Architect left the cosmos to its own devices, free to evolve by chance and necessity. Eventually, creatures arose from the primeval slime and built temples to praise this Supreme Being. This is the *Creator* or *Divine Providence* that the American Declaration of Independence speaks of. Thomas Jefferson and Benjamin Franklin were deists, as a belief in such a naturalistic God is called.

Science gives a valid description of the way things interact with each other and how they change from one form into another. That galaxies, cars, billiard balls, and subatomic particles act in a regular manner that can be captured by mathematics, and that can therefore be predicted, is nothing short of amazing. Indeed, some physicists and mathematicians — most famously Albert Einstein — believe in such a creator precisely because of this "miraculous" state of affairs. It is not difficult to imagine a universe so complex that it is incomprehensible. But the deist's God created a universe that is not only hospitable to life, but also so predictable that its regularity can be apprehended by the human mind.

Yet we search in vain for direct empirical evidence for such an eternal being above and beyond natural forces. God leaves no residue in our test tubes nor tracks in our bubble chambers. God does not reveal himself through logic, either. As the astronomer-philosopher Immanuel Kant argued, all proofs of the existence of God are flawed. No chain of bullet-proof arguments leads from unassailable propositions to the firm conclusion that God must, of necessity, exist. (The opposite is also true — it can't be proved that there is no God.) Again, Wittgenstein puts it sparingly: "God does not reveal himself *in* the world."

The early 1970s saw a new twist to this debate — namely, the *anthropic principle*, the observation that the universe is ever so friendly to stable, self-replicating biochemical systems. If the physical constants and parameters that govern the cosmos were slightly different, complex molecules, and therefore life, could not exist. In that sense, "anthropic" is a misnomer, for the principle does not refer to human life; rather, it should be called the "biotropic" or, perhaps, the "biophilic" principle.

Take Newton's law of gravity and Coulomb's law, which governs the way electrically charged particles attract and repel each other. Both laws have the same form, stating that the forces decay with the square of the distance between any two particles. Only the constant in front of the quadratic decay term differs. Intriguingly, the attraction between two opposite charges must be exactly 10,000 trillion, trillion, trillion times stronger than their mutual gravitational attraction in order for life as we know it to form. A itsy-bitsy more or less and we would not be around.

Another cosmic constraint is that the sum of all positively charged particles in the universe must equal the sum of all negatively charged particles; otherwise, electromagnetism would dominate gravity, and stars, galaxies, and planets couldn't form. The number of electrons must equal the number of protons to within one part in a trillion, trillion, trillion. If the strong nuclear force were slightly stronger or weaker than it actually is, either nothing but hydrogen or no elements heavier than iron would exist. If the universe had expanded too rapidly, protons and neutrons couldn't have bonded into atomic nuclei. If the initial expansion of the universe had been ever so slightly slower, the fiery brew that made up the early universe would have been too hot for nuclei to have formed. In short, an amazing number of "coincidences" had to occur to give rise to a universe that was stable for a sufficiently long time and diverse enough in chemical elements to support complex, carbon-based life forms.

Some argue that the anthropic principle is tautological: If the universe had not been friendly to life, we wouldn't be around to ponder its existence. This assumes that there is an uncountable number—in time or in space—of parallel universes that are inhospitable to life and that we happen to be in the one that is conducive to life. The trouble is that we don't know about these other universes—they have never been observed. Maybe we live in a multiverse, containing an infinite number of noninteracting and nonobservable universes. Possibly. But assuming an uncountable number of worlds is a very strong assumption and is as *ad hoc* as the hypothesis of a Supreme Architect who rigged the laws of physics to facilitate the formation of life.

The lively debate triggered by the anthropic principle shows no signs of settling down.

What remains is neither empirical knowledge nor logical certainty, but faith. Some, such as the physicists Stephen Hawking and Leonard Mlodinow, express their faith that a yet-to-be-proven theory of physics called M-theory will demonstrate why the universe has to exist the way it does. Others find this a questionable promissory note at best and have faith in different principles.

Theism, or God as the Interventionist

What power does God have? Can the Supreme Being influence the course of events in his creation? After all, people pray in the expectation that God will listen to them—provided that their intentions are pure and their belief sincere—and will intercede on their behalf to cure a sick child, steady a rocky marriage, or bless a new business. If God is unable to do any of these things, why bother? (I'm not concerned here with the possible beneficial psychological effects of praying, such as relief from anxiety. I'm after bigger game.)

Theism is the belief in an activist God who intervenes in the universe. Is Theism compatible with science? When something outside the universe, not subject to natural law, causes something inside the universe to happen, people speak of a miracle. So, the question needs to be rephrased as, are miracles compatible with science? The answer is an unambiguous no.

Take Jesus' first public action (according to the New Testament), turning water into wine at a wedding in Canaan. This runs counter to a fundamental principle, the conservation of mass–energy. The aromatic and ether molecules making up the wine had to come from somewhere. Water molecules can be converted into carbon and the other elements and molecules that constitute wine, but this is a feat of nuclear synthesis that requires prodigious amounts of energy. Nothing like that was reported.

Every time this conservation principle is tested, it is found to be sound, from the infinitesimally small to the unimaginably large. It is, therefore, extremely unlikely that the miracle in Canaan took place.

Scientists are guided by a heuristic principle of deductive reasoning called *Occam's razor*. Named after the fourteenth-century English friar and logician William of Ockham, it states that of two equally good explanations for a phenomenon, the simpler one is more likely to be true. A more convoluted explanation is less likely than a more parsimonious one. It is not a logical principle but a working rule.

When reconstructing an anomalous event, a murder or an airplane crash, investigators can never determine with absolute certainty what happened. But Occam's principle narrows the options. Occam's razor slashes hither, eliminating the unknown assailant with no apparent motive who left no physical trace but who, so claims the defense attorney, was responsible for the murder. Occam's razor slashes thither, decimating the theory of a secret government conspiracy that brought down the airplane but that necessitates an unlikely chain of events and the active participation of many people. Occam's razor is an invaluable tool, eliminating superfluous entities from consideration.

The possibility of a Supreme Being turning water into wine is so outlandish that it can be rejected using Occam's razor. It is far more likely that something else, obeying the laws of physics, was the cause. Maybe the wedding organizers discovered long-forgotten flasks of wine in the basement. Or a guest brought the wine as a gift. Or the story was made up to cement Jesus' reputation as the true Messiah. Remember Sherlock Holmes' advice: "When you have eliminated the impossible, whatever remains, however improbable, must be the truth."

Miracles are not in the cards. The fabric of everyday reality is woven too tightly for it to be pulled asunder by extranatural forces. I'm afraid that God is an absentee cosmic landlord. If we want things to happen down here, we had better take care of them ourselves. Nobody else is going to do it for us.

Can Revelation and Scripture Be Helpful?

Traditionally, the most important source of knowledge about the transcendental is revelation—the direct, first-hand experience of God. Saul's encounter with the living God on the road to Damascus turned him from a persecutor of the early acolytes of Jesus into the apostle Paul, Christianity's greatest missionary. The seventeenth-century French mathematician, physicist, and philosopher Blaise Pascal likewise experienced God in this manner: A description of his searing experience was found on a parchment sewn into the lining of his coat. The writings of saints and mystics from all religious traditions contain encounters with the Absolute and the feeling of oneness with the universe.

If I had experienced God in this manner, if I would have seen the burning bush and felt some manifestation of the *Mysterium tremendum*, I would not be writing these lines. I wouldn't have to resort to inadequate reason to figure things out.

Because I only have reason to fall back on, I admit to skepticism when considering the ontological (but not the psychological) validity of such life-changing experiences. As a husband, father, son, brother, friend, lover, colleague, scientist, citizen, and avid reader of history, I continue to be amazed by the ability of highly educated and intelligent people to fool themselves. You and I are convinced that our motives are noble, that we are smarter than most, that the opposite sex finds us attractive.

Nobody is immune from self-deception and self-delusion. We all have intricate, subliminal defense mechanisms that allow us to retain beliefs that are dear to us, despite contravening facts. September 11, the Iraq debacle, and the Lehman Brothers' bankruptcy vividly demonstrate that the "elite" suffer from these failures of common sense as much as anybody else. A Caltech colleague, the physicist Richard Feynman, said "The first principle is that you must not fool yourself, and you are the easiest person to fool." Our pathological propensity for interpreting any event in the light of what we want to believe is exactly why double-blind experimental protocols are so essential in science and medicine. They root out the experimentalist's hidden biases, which otherwise contaminate the results.

Given these uncomfortable facts about human nature, I am doubtful that intense religious experience, although no doubt genuinely felt, reveals anything about the actual existence of God. Too much is at stake for people to be objective. I do not deny such experiences, but I am wary of their interpretation. I stand ready with sodden blankets and the cold water of reason.

I am equally skeptical when turning to Holy Scripture, another traditional source of religious thought and teaching. The idea that the experiences and thoughts of men who lived thousands of years ago are relevant to our understanding of the universe and our place in it strikes me as quaint. The books of the Bible were written before the true age and extent of the cosmos were even remotely imagined, before the evolutionary bonds between humans and animals were understood, before the brain was identified as the seat of the mind (neither the Old nor the New Testament mentions the brain a single time).

Moreover, the observation that different societies and cultures have foundational texts and traditions that are quite at odds with each other, differences that some followers are willing to kill and to die for, does not increase my confidence in these received "truths." (What strange gods, rituals, and beliefs will join this pantheon if we ever find civilizations on distant stars? Can they, too, attain salvation? Did Jesus die for them as well?) Given this marketplace of religions, on what basis should I choose one over the others? For many years I, like the vast majority of people, believed what my parents believed. But that is not a truly informed choice.

The Old and New Testaments, the Koran, and other religious texts are poetic, inspirational, and insightful about enduring human needs and desires. They provide the ethical foundations that have guided the faithful for millennia. Holy writ reminds us, again and again, that each individual is part of something larger, part of a larger community of believers, and part of creation. Contemporary cultural, political, and social life revolves around the golden calf of greed and consumption. Wars and conflicts, stock market crashes, and environmental degradation and shortages of water and oil remind us that we neglect these essential truths at our peril. But as we learn more about the universe, the relevance of these sacred texts to the contemporary world lessens.

Let me give you a personal example of how the scientific insights discussed in this book have made a concrete difference in my life. As I explained earlier, many—and possibly all—animal species have subjective, phenomenal feelings. They experience pleasure and pain, are happy and sad. In light of this knowledge, how can we justify raising animals under atrocious, industrial conditions, far removed from their natural habitats, in order to eat them? How can we breed sentient beings for their flesh? How can we justify locking up young calves in tight, closed boxes where they can't turn around or lie down and deprive them of any social contact for the duration of their short life just so we can eat their white and tender flesh? This is particularly barbaric today, when nutritious, good-tasting, cheap—and more healthful—alternatives to meat are readily available. Yet, it was difficult for me to follow through on this intellectual insight with action—the taste of flesh is very deeply ingrained in our cuisine and palates. Then, in 2004, Susan Blackmore, an intrepid British psychologist with rainbow-colored hair, interviewed me for a book of hers. I had just concluded a riff on mouse consciousness with a plea to not kill mice thoughtlessly, as many researchers who work with them do, when Susan asked me, out of the blue, whether I ate meat. We looked at each other for a while, silently, until I sighed to cover up my embarrassment at having been revealed a hypocrite. This incident really bothered me.

When a year later my beloved Nosy died, I was moved to act. Of the six dogs I have lived with, I grew fondest of this ever-so-savvy, playful, and intensely curious black German shepherd. When she passed away, I was distraught; I still dream about her today. I asked myself that night, as she lay dying in my arms, how could I cry over her but happily eat the flesh of lambs and pigs? Their intelligence and brains are not that different from those of dogs. From that night on, I stopped eating mammals and birds, though somewhat inconsistently, I still consume fish.

None of the Ten Commandments teaches us to avoid eating the flesh of sentient creatures. None of them instructs us to take care of planet Earth. The Decalogue is not helpful in making end-of-life decisions or dealing with reproductive cloning. We need a new set of commandments, one appropriate for our times, as is forcefully advocated by the philosopher and ethicist Peter Singer, one of the founders of the modern animal rights movement.

Et in Arcadia Ego

My upbringing left me with a yearning for the absolute and with the recognition that the numinous can be found in all things—the howling of a dog, the sight of the star-studded heavens, the contemplation of the periodic table, or the pain of ice-cold hands during a windy climb.

On occasion, I have encountered a dark side to such lucubrations. As a teenager lying in bed at night, I would strive to grasp eternity. What does it feel like for time to be going on forever? What does it mean to be dead forever? Not just dead for a century, or a millennium, not just for a long time or a really long time, but forever and ever. The conceptual artist Roman Opalka sought to plumb infinity, to capture it—the steady, incomprehensible progression of one number after another, from one to infinity, on and on. Painting this endless stream of numbers during the last forty-five years of his life was Opalka's way of dealing with the dizzying and terrifying notion of the infinity stretching in front of us.

Yet, I never worried about my own demise. Like many young men in pursuit of the extreme—whether it be climbing, motorcycle racing, finance, or war—I didn't think about the end. It wasn't really going to happen to me. Not even the death of my daughter Elisabeth shocked me out of this blessed complacency.

It was only in my early forties that I truly realized death was going to come for me, too. I have told the story in the opening pages of chapter 6. One night, my unconscious rebelled. I woke up, and abstract knowledge had turned to gut-wrenching certainty: I was actually going to die!

I pondered the significance of my personal annihilation for the next several months, facing down an existentialist abyss of oblivion and meaninglessness within me. Eventually, through some unconscious process of recalibration, I returned to my basic attitude that all is as it should be. There is no other way I can describe it: no mountaintop conversion or flash of deep insight, but a sentiment that suffuses my life. I wake up each morning to find myself in a world full of mystery and beauty. And I am profoundly thankful for the wonder of it all.

Here I am, a highly organized pattern of mass and energy, one of seven billion, insignificant in any objective accounting of the world. And in a short while I will cease to exist. What am I to the universe? Practically nothing. Yet the certainty of my death makes my life more significant. My joy in life, in my children, my love of dogs, running and climbing, books and music, the cobalt blue sky, are meaningful *because* I will come to an end. And that is as it should be. I do not know what will come afterward, if there is an afterward in the usual sense of the word, but whatever it is, I know in my bones that everything is for the best.

This sentiment is tied in with my overall sunny and optimistic disposition, which is largely determined by genetic factors and was amplified by the benign circumstances in which I grew up. I can't take credit for either.

Edith is the strong, centered, and responsible woman who kept me grounded for close to three decades. She enabled me to develop fully as a professor and a scientist. She put her career on hold for many years to raise our children into the healthy, smart, resourceful, responsible, and beautiful adults that they are today. This meant that I could read and sing them to sleep, travel to foreign countries and hike, camp, and river raft with them, help with their homework and school projects, and indulge in all the other pleasures of fatherhood without any significant sacrifices to my professional life.

And we enjoyed the company of a fluctuating number of big, hairy, boisterous dogs—Trixie, Nosy, Bella, and Falko. Next to children, they are the best things in life.

Together with a handful of colleagues, I inaugurated and directed two science summer schools: one on computational neuroscience at the Marine Biological Laboratory at Woods Hole, Massachusetts, on the Atlantic shore, and another one on neuromorphic engineering (what engineers can learn from neurobiology) in Telluride, Colorado, in the Rocky Mountains. Both remain popular. Each summer, our family spent four intense weeks in these glorious places. This was the happiest period of my life.

Those halcyon days came to an end when my son and daughter left for college. I missed them more than anything else. To fill that enormous void and to keep my energy channeled, I took up rock climbing in the Sierra Nevada and Yosemite Valley, long-distance trail running in the local mountains, marathons in Death Valley, and so on—anything challenging to combat my growing restlessness. I was suffering from acute empty nest syndrome.

Then Francis left my life. I was with him in his home study when his oncologist called, confirming that his colon cancer had returned with a vengeance. He stared off into the distance for a minute or two, and then returned to our reading. This diagnosis was discussed with Odile during lunch, but that was the extent of it for that day. Of course, I wasn't reading his dark thoughts during that night. But I did recall an earlier conversation in which he confessed to me that his own, not too distant, end made him sad but that he was resolved to not waste any remaining time by fruitless ruminations and ponderings nor by chasing high-risk experimental therapies. Here I saw him practice this resolution. What mind control! What composure!

A few months later, suffering from the nauseating effects of chemotherapy that failed to stop the spread of the cancer, he put down the telephone in the neighboring room and shuffled past me on the way to the bathroom. When he returned to resume the phone conversation, he dryly remarked in passing, "Now I can truly tell them that their idea made me throw up." (Somebody was trying to convince Francis to sign off on the creation of a bobblehead of him.) With a view of the inevitable, Francis gave me a life-sized, black-and-white photograph of him, as I knew him. Sitting in a wicker chair, he ironically gazes at the viewer with a twinkle in his eyes. Signed "For Christof—Francis—Keeping an eye on you," it watches over me in my office.

In summer 2004, Francis phoned me on the way to the hospital to tell me that his emendations to our last manuscript, on the function of the claustrum, a sheet-like structure beneath the cortex, would be delayed. Yet he kept on working, dictating corrections to his secretary from the clinic. Two days later he died. Odile recounted how on his deathbed, Francis hallucinated arguing with me about rapidly firing claustrum neurons and their connection to consciousness, a scientist to the bitter end. He was my mentor, my close intellectual companion, and my hero for the unflinching manner in which he dealt with aging and mortality. His absence left a gaping hole in my life.

My father had passed away in the opening weeks of the third millennium, leaving me without an elder mentor to turn to for guidance and support. My disquiet was compounded, paradoxically, by the successful publication of *The Quest for Consciousness*. I had worked hard over many years toward that goal, which at times appeared very distant. Now that the race was run, I felt listless, bereft of a clear mission. I needed the challenge of an Annapurna in my life.

Precipitated by these cumulative departures, I grew estranged from my wife and left. It is easy enough to state this matter-of-factly, but these few words encompass a degree of misery, distress, pain, and anger over a protracted period that is impossible to put to paper. (Watch Ingmar Bergman's cimmerian masterpiece *Scenes from a Marriage* to understand what I mean.) I went through a searing crisis, experiencing first-hand the power of the unconscious to shape feelings and actions in a way that escapes conscious insight. Once those forces were unleashed, I was unable to master them. Or perhaps I was unwilling to master them. There is a reason that Dante consigns sinners who "made reason subject to desire" to the Second Circle of Hell. It was, without a doubt, the grim, low point of my life. Yet something was compelling me onward.

Spinoza coined a beautiful expression, *sub specie aeternitatis*, literally, "under the form of eternity." This is the remote viewpoint. Look down onto the Milky Way from far above its central black hole. You see a swirling disk of hundreds of billion of stars, many of them surrounded by tiny, dark associates. Some of these planets harbor life. On one of them, semi-intelligent, violent, and social primates furiously couple and uncouple. They endow this frenetic, anthill activity with great cosmic importance. These pairings last but the blink of an eye, the flash of a firefly, the flight of an arrow, compared with the time it takes the majestic galactic wheel to complete one rotation.

My anguish begins to recede in significance when viewed in this celestial light of deep time. My tribulations are not meaningless—I am no nihilist—but they should not, and will not, overwhelm my life. Having lost my central sun, I am a solitary planet now, wandering in the silent spaces between the stars. I am slowly regaining some measure of that inner peace, the equanimity, what the Epicureans called ataraxia, that I had for so long.

To come to terms with my actions, I studied what science knows about voluntary acts and free will, which explains the genesis of chapter 7. What I took from my readings is that I am less free than I feel I am. Myriad events and predispositions influence me. Yet I can't hide behind biological urges or anonymous social forces. I must act as if "I" am fully responsible, for otherwise all meaning would be leached from this word and from the notions of good and bad actions.

One night, in the midst of my crisis, I emptied a bottle of Barolo while watching the fantasy action movie *The Highlander* and felt in need of some symbolic gesture. At midnight I decided to run to the top of Mount Wilson, which rises more than five thousand feet above Pasadena. After an hour stumbling around with my headlamp and becoming nauseated, I realized that I was being stupid and turned back—but not before shouting into the dark the closing line of the poem *Invictus*: "I am the master of my fate, I am the captain of my soul." This is a perhaps overenthusiastic expression of my position on the question of free will: For better or worse, I am the principal actor of my life.

Now that you've arrived at the coda of this autobiographical chapter, I can confess what has become obvious by this point. I was driven to write this book for a trifecta of reasons—to describe my quest for the material roots of consciousness, to come to terms with my personal failings, and to bring my search for a unifying view of the universe and my role in it that does justice to both chance and necessity to a satisfactory conclusion.

Nailing My Colors to the Mast

This is the end of my story. I'm optimistic that science is poised fully to comprehend the mind–body problem. To paraphrase language from Corinthians, "For now we see through a laboratory, darkly, but then shall we know."

I do believe that some deep and elemental organizing principle created the universe and set it in motion for a purpose I cannot comprehend. I grew up calling this entity God. It is much closer to Spinoza's God than to the God of Michelangelo's painting. The mystic Angelus Silesius, a contemporary of Descartes, captures the paradoxical essence of the self-caused Prime Mover as "*Gott ist ein lauter Nichts, ihn rührt kein Nun noch Hier*" (God is a lucent nothing, no Now nor Here can touch him).

The pioneering generation of stars had to die in spectacular supernova fashion to seed space with the heavier elements necessary for the second act of creation—the rise of self-replicating bags of chemicals on a rocky planet orbiting a young star at just the right distance. The competitive pressures of natural selection triggered the third act of creation—the accession of creatures endowed with sentience, with subjective states. As the complexity of their nervous systems grew to staggering proportions, some of these creatures evolved the ability to think about themselves and to contemplate the splendidly beautiful and terrifyingly cruel world around them.

The rise of sentient life within time's wide circuit was inevitable. Teilhard de Chardin is correct in his view that islands within the universe—if not the whole cosmos—are evolving toward ever-greater complexity and self-knowledge. I am not saying that Earth had to bear life or that bipedal, big-brained primates had to walk the African grasslands. But I do believe that the laws of physics overwhelmingly favored the emergence of consciousness. The universe is a work in progress. Such a belief evokes jeremiads from many biologists and philosophers, but the evidence from cosmology, biology, and history is compelling.

Spiritual traditions encourage us to reach out to our fellow travelers on the river of time. More than most secular ideologies, religions emphasize the common bond among people: love thy neighbor as thyself. Religious sentiments, as expressed through music, literature, architecture, and the visual arts, bring out some of what is best in humankind. Yet collectively, they are only of limited use in making sense of the puzzle of our existence. The only certain answers come from science. What I find most appealing from an intellectual and ethical point of view are certain strands of Buddhism. But that is a topic for another book.

I am saddened by the loss of my religious belief, like leaving forever the comfort of my childhood home, suffused with a warm glow and fond memories. I still have feelings of awe when entering a high-vaulted cathedral or listening to Bach's *St. Matthew Passion*. Nor can I escape the emotional thrall, the splendor and pageantry, of high Mass. But my loss of faith is an inescapable part of growing up, of maturing and seeing the world as it is.

I'm cast out into the universe, a glorious, strange, scary, and often lonely place. I strive to discern through its noisy manifestations—its people, dogs, trees, mountains, and stars—the eternal Music of the Spheres.

When all is said and done, I am left with a deep and abiding sense of wonder. Echoing across more than two thousand years, the unknown scribe of the Dead Sea Scrolls, living in a tiny community in the Judean desert, expressed it well. Let his psalm close my book:

So walk I on uplands unbounded
and know that there is hope
for that which thou didst mold out of dust
to have consort with things eternal.

Notes

In 2004, I published *The Quest for Consciousness*. Summarizing the approach Francis Crick and I took, it describes the neurobiological circuits and the psychological processes critical to consciousness. It contains hundreds of footnotes spread over four hundred single-spaced pages and close to a thousand references to the scholarly literature. *Confessions* treads more lightly. If you want to find out about the techniques and experiments mentioned in this book, please consult *Quest*, the edited volume by Laureys and Tononi (2009), the crowd-sourced and up-to-date *Wikipedia*, or the sparse notes that follow. I also write a regular column for *Scientific American Mind*, called *Consciousness Redux*, which features contemporary consciousness research.

Chapter 1

The short biography of Francis Crick by Ridley (2006) does an excellent job of portraying Francis's character. The tome by Olby (2009) is much more magisterial, with expositions of Francis's scientific contributions. Olby's penultimate chapter covers Francis's and my collaboration.

Consult Chalmers (1996) for the origin and meaning of the term *Hard Problem*.

Chapter 2

Koch and Segev (2000) summarize the biophysics of single neurons.

Mann and Paulsen (2010) describe the effects that the local field potential, generated by tens of thousands of neurons, has on the firing of these nerve cells. The experiments by Anastassiou and Perin (Anastassiou et al., 2011), directly show how the firing of neurons can be synchronized by weak external electric fields.

The economy also fails to obey conservation laws: A company may be worth billions of dollars one day but only millions the next, even though nothing has changed on the ground. The same people work in the same buildings with the same infrastructure. Where has the money gone to? The answer is that the market's belief in the future of the company, its expectation, has abruptly evaporated, and with it the market's evaluation. Unlike energy, money can be created and destroyed.

Chapter 3

Although dated, Francis's terse introduction to consciousness and its biological basis remains masterful (Crick, 1995).

The synthesis of a biological organism by Craig Venter is described in Gibson and others (2010).

The Tyndall quote is from his presidential address to the mathematical and physical section of the British Association for the Advancement of Science in 1868, entitled "Scientific Materialism" (Tyndall, 1901).

The Tannhäuser Gate quote is, of course, from the final scene in *Blade Runner*. Directed by Ridley Scott, it is the greatest science fiction movie ever. It is loosely based on *Do Androids Dream of Electric Sheep?* mentioned a few lines later. This remarkable novel, written in 1968 by Philip K. Dick, prefigures the "uncanny valley" effect, the psychological observation that a robot or computer-animated graphical replica of a human that is almost but not quite perfect evokes a feeling of revulsion.

The Huxley quote comes from a remarkable speech he delivered in 1884 to the same British association that listened to Tyndall sixteen years earlier. Huxley took issue with Descartes' belief that animals were mere machines or automata, devoid of conscious sensation. He assumed that for reasons of biological continuity, some animals shared facets of consciousness with humans, but was at a loss when it came to the function of consciousness.

The best introduction to animal consciousness is the short *Through Our Eyes Only?* by Dawkins (1998). Alternatively, see the encyclopedic Griffin (2001). Edelman and Seth (2009) focus on consciousness in birds and cephalopods.

A modern, concise introduction to neuroanatomy is Swanson (2012).

The Krakauer quote is from his superb collection of essays on alpinism from 1990.

Watching *The Good, The Bad and the Ugly* inside a magnetic scanner was reported in Hasson et al. (2004).

The inactivation of regions relating to self is dealt with in Goldberg, Harel, and Malach (2006).

Chapter 4

Our thinking on the neuronal correlates of consciousness evolved (Crick and Koch, 1990, 1995, 1998, 2003). The philosopher David Chalmers eloquently summarized the metaphysical and conceptual assumptions underlying the idea of correlates of consciousness in Chalmers (2000; see also the essay by Block, 1996). Tononi and Koch (2008) provide an update of relevant experimental studies.

Rauschecker and colleagues (2011) electrically stimulate the surface of visual cortex to elicit visual motion percepts in a neurosurgical setting.

Macknik and others (2008) are among the first who pointed to the rich sources of insight from stage magic for psychology and neuroscience.

The continuous flash suppression (CFS) technique for hiding objects in plain sight for fractions of a minute or longer was developed in Tsuchiya and Koch (2005).

A more recent and elegant example of using CFS to charter the unconscious mind is Mudrik and colleagues (2011). A compendium of masking methods can be found in Kim and Blake (2005). Jiang and others (2006) asked volunteers to look at invisible pictures of nude men and women, and Haynes and Rees (2005) carried out fMRI on the brains of volunteers looking at invisible gratings. The visual word form area and its relationship to reading is described by McCandliss, Cohen, and Dehaene (2003).

Logothetis (2008) well summarizes the promise and the limit of fMRI in deciphering the underlying differential neuronal responses. His classical studies on the neuronal basis of binocular rivalry are surveyed in Logothetis (1998) and Leopold and Logothetis (1999).

A Japanese-German team (Watanabe et al. 2011) dissociated visual attention and consciousness—as assayed via stimulus visibility—in human primary visual cortex. Whether or not subjects see what they are looking at makes very little difference to the hemodynamic signal in V1, while attention strongly modulates it.

The bulk of the research I discuss here on dissociation between neural activity in primary visual, auditory, and somatosensory cortices and conscious seeing is referenced in chapter 6 in *Quest*.

Support for a functioning connection from the front of cortex to the back to sustain consciousness in severely disabled patients is provided by Boly et al. (2011). See also Figure 1 in Crick and Koch (1995).

The philosopher Ned Block is enormously influential in the debate concerning the relationship between attention and consciousness (Block, 2007). van Boxtel, Tsuchiya, and

Koch (2010) survey the myriad experiments teasing selective visual attention apart from visual consciousness.

Chapter 5

Gallant et al. (2000) describe patient A.R.

The latest book by my favorite neurologist, Oliver Sacks (2011), contains evocative descriptions of face-blindness and other neurologic deficits. Sacks is an avid witness of the human condition. In examining how people deal with disease, he demonstrates how they, and we, can gain wisdom about life.

One of the pioneers of cortical electrophysiology, Semir Zeki, coined the term *essential node* in Zeki (2001).

The scientific legacy of the amnesic patient H.M. is summarized in Squire (2009).

Quian Quiroga et al. (2005, 2009) discovered concept cells in the human medial temporal lobe that respond to pictures, text, and voices of celebrities or familiar persons. These are closely related to so-called *Grandmother neurons* (Quian Quiroga et al. 2008). Cerf and others (2010) exploit computer feedback to enable patients to control these neurons by their thought.

Owen et al. (2006) and Monti et al. (2010) garnered international attention when they reported the detection of consciousness in a few vegetative patients using a magnetic scanner.

Parvizi and Damasio (2001) explore the link between the forty or more brainstem nuclei and consciousness.

Laureys (2005) maps out our astonishingly dynamic ideas about the link between death, the brain, and consciousness. Schiff (2010) is a neurologist who specializes in the recovery of consciousness after massive brain trauma.

Chapter 6

Vast amounts of nonsense continues to be put out about the unconscious. However, there has been a revival of solid, empirical studies of the nonconscious under carefully controlled conditions. Hassin, Uleman, and Bargh (2006) provide a book-length treatment of some of the best post-Freud research, while Berlin (2011) focuses on the knowns and the unknowns of the neurobiology of the unconscious.

Jeannerod's research is described in his 1997 book.

Denial-of-death mechanisms are considered as possible evolutionary drivers by Varki's (2009) letter.

Experiments demonstrating that your eyes can resolve finer details than you can see were carried out by Bridgeman and others (1979) and by Goodale and colleagues (1986). Logan and Crump (2009) show that the typist's hands know things that the typist's mind doesn't. The two visual streams theory of conscious perception and unconscious visuo-motor action is reviewed in Goodale and Milner (2004).

The MIT historian John Dower (2010) exhaustively analyzed the structural similarities and differences between Pearl Harbor and September 11, along with other pathologies of institutional decision making.

The priming experiments are taken from Bargh, Chen, and Burrows (1996). Johansson and colleagues (2005) asked men and women to chose which of two women look more attractive and then swapped their images without most of the volunteers catching on.

Chapter 7

The reflections on the physics of free will were influenced by my reading of Sompolinksy (2005).

Sussman and Wisdom (1988) prove that Pluto's orbit is chaotic.

The extent to which fruit flies display true randomness is studied by Maye et al. (2007).

Turner (1999) links quantum fluctuations in the early universe with today's distributions of galaxies in the sky. The physicist Jordan (1938) proposed the still popular—in some quarters—quantum amplifier theory linking elementary particle physics to free will. Koch and Hepp (2011) discuss the possible relevance of quantum mechanics to the brain. Collini et al. (2010) provide compelling evidence of electronic coherence in photosynthetic proteins at room temperatures.

The original report of brain activity prior to the feeling of having willed the action is by Libet and colleagues (1983). A brain imaging variant of the original experiment by Soon and colleagues (2008) caught the public's eye.

There is a growing literature on the neuropsychology of free will (Haggard, 2008). Murphy, Ellis, and O'Connor (2009) edited a book trying to resolve some of the tensions between traditional notions of free will, grounded in theology and everyday experience, and those from modern psychology and biology.

The morbid story of how *T. gondii* hitchhike a ride in the brain of the rat, forcing it to change its behavior so it is more likely to be eaten by cats, is told by Vyas et al. (2007). The possible ramifications for human culture of brain infections by this parasite are analyzed by Lafferty (2006).

Wegner (2003) superbly describes the psychology of voluntary actions in normal life and under pathologic conditions.

Two neurosurgical studies electrically stimulated the brain to initiate a "voluntary" action (Fried et al. 1991; Desmurget et al. 2009).

Chapter 8

Baars's (2002) book describes his *Global Workspace* model of consciousness. Dehaene and Changeux (2011) review imaging and physiologic data in support of a neuronal implementation of the global workspace model.

Chalmers's ideas on information theory and consciousness are sketched out in an appendix to his 1996 book.

The easiest introduction to Tononi's theories and thoughts is his 2008 manifesto. I warmly recommend the highly literary and novelistic treatment of the relevant facts and the theory by Giulio himself in his *Phi* (2012), in which Francis Crick, *Alan Turing*, and Galileo Galilei undertake a voyage of discovery through the Baroque. For the actual mathematical calculus, consult Balduzzi and Tononi (2008, 2009). Barrett and Seth (2011) derive heuristic to compute integrated information.

Consult chapter 17 of my *Quest* on the background to split-brain patients. A novelistic treament of life with split hemispheres is Lem (1987).

Of the 86 billion neurons in the human brain, an amazing 69 billion are in the cerebellum and 16 billion in cortex (Herculano-Houzel, 2009). That is, about four of every five brain cells are cerebellar granule neurons with their stereotyped four stubby dendrites. The remaining one billion neurons are in the thalamus, basal ganglia, midbrain, brain stem, and elsewhere. Yet what is remarkable are the few cognitive deficits in people born without cerebellum—a rare occurrence—or in patients that lose part of the cerebellum due to stroke or other trauma. The main deficits are ataxia, slurred speech, and unsteady gait (Lemon and Edgley, 2010).

Evolving *animats*—caricatures of creatures that live inside a computer—over tens of thousands of generations demonstrates that the more adapted these become to their simulated environment, the higher their integrated information (Edlund and colleagues, 2011).

Koch and Tononi (2008) treat the prospects of machine consciousness in light of the integrated information theory of consciousness. We propose an image-based test of what it means for a computer to consciously see pictures in Koch and Tononi (2011).

Chapter 9

The combo of TMS-EEG to test the breakdown of the conscious mind in sleep is described by Massimini et al. (2005). The extension of this to persistent vegetative state and minimal conscious patients is found in Rosanova et al. (2012).

Neuroanatomists still do not know whether or not the biggest mammal, the blue whale, with 17 pounds (8,000 grams) of brain, has more neurons than humans do. Bigger brains does not necessarily imply more neurons; it would be quite a shock, though, if cetaceans and elephants had more brain cells than humans. For a thorough discussion of brain size and number of neurons, see Herculano-Houzel (2009).

A short primer on neuronal cell types is Masland (2004).

The use of optogenetics is exploding, with hundreds of laboratories manipulating genetically identifiable cell populations at a location and time of their own choosing. This is remarkable, given that the original paper pioneering the usage of channelrho-dopsin-2 to drive neural activity was published in Boyden and colleagues (2005). For my money, the three most elegant optogenetic experiments linking neural circuits in a causal manner to mouse behavior are Adamantidis et al. (2007; this is the orexin study I describe), Gradinaru et al. (2009), and Lin et al. (2011). A compendium of current techniques can be found in Gradinaru et al. (2010).

The Allen Brain Atlas of the mouse is comprehensively described in Lein et al. (2007) and can be found online.

Dehaene et al. (2001) measures the fMRI response of volunteers looking at visible and invisible words.

Chapter 10

My embarrassing interview can be found in Blackmore (2006).

The literature on the relationship between science and religion is vast. I found the 2008 book by Hans Küng, a famed liberal theologian, helpful.

The portrait of Francis hanging in my office was taken by Mariana Cook for her collection *Faces of Science* (2005).

Francis literally worked on the claustrum paper until his dying breath. It was published as Crick and Koch (2005).

The 1994 book by the philosopher Peter Singer on the inadequacy of traditional ethics to address modern challenges to life and death is very insightful.

Despite the slightly bombastic ending, nothing in my book should be construed as taking me or my life too seriously. The walls of the tattoo parlor where I engraved a cortical microcircuit from Ramón y Cajal onto my left arm are graced with an admonition that serves as alternative ending:

Life's journey is not to arrive at the grave safely, in a well preserved body, but rather to skid in sideways, totally worn out, shouting, "Holy mackerel . . . what a ride!"

References

Adamantidis, A. R., Zhang, F., Aravanis, A. M., Deisseroth, K., & de Lecea, L. (2007). Neural substrates of awakening probed with optogenetic control of hypocretin neurons. *Nature, 450*, 420–424.

Anastassiou, C. A., Perin, R., Markram, H., & Koch, C. (2011). Ephaptic coupling of cortical neurons. *Nature Neuroscience 14*, 217–223.

Baars, B. J. (2002). The conscious access hypothesis: Origins and recent evidence. *Trends in Cognitive Sciences, 6*, 47–52.

Balduzzi, D., & Tononi, G. (2008). Integrated information in discrete dynamical systems: Motivation and theoretical framework. *PLoS Computational Biology, 4*, e1000091.

Balduzzi, D., & Tononi, G. (2009). Qualia: The geometry of integrated information. *PLoS Computational Biology, 5*, e1000462.

Bargh, J. A., Chen, M., & Burrows, L. (1996). Automaticity of social behavior: Direct effects of trait construct and stereotype activation on action. *Journal of Personality and Social Psychology, 71*, 230–244.

Barrett, A. B., & Seth, A. K. (2011). Practical measures of integrated information for time-series data. *PLoS Computational Biology, 7*, e1001052.

Berlin, H. A. (2011) The neural basis of the dynamic unconscious. *Neuro-psychoanalysis, 13*, 5–31.

Blackmore, S. (2006). *Conversations on Consciousness: What the Best Minds Think about the Brain, Free Will, and What It Means to Be Human.* New York: Oxford University Press.

Block, N. (1996). How can we find the neural correlate of consciousness? *Trends in Neurosciences, 19*, 456–459.

Block, N. (2007). Consciousness, accessibility, and the mesh between psychology and neuroscience. *Behavioral and Brain Sciences, 30*, 481–499, discussion 499–548.

Boly, M., & Associates. (2011). Preserved feedforward but impaired top-down processes in the vegetative state. *Science, 332*, 858–862.

Boyden, E. S., Zhang, F., Bamberg, E., Nagel, G., & Deisseroth, K. (2005). Millisecond-timescale, genetically targeted optical control of neural activity. *Nature Neuroscience, 8*, 1263–1268.

Bridgeman, B., Lewis, S., Heit, G., & Nagle, M. (1979). Relation between cognitive and motor-oriented systems of visual position perception. *Journal of Experimental Psychology. Human Perception and Performance, 5*, 692–700.

Cerf, M., Thiruvengadam, N., Mormann, F., Kraskov, A., Quian Quiroga, R., Koch, C., et al. (2010). On-line, voluntary control of human temporal lobe neurons. *Nature, 467*, 1104–1108.

Chalmers, D. J. (1996). *The Conscious Mind: In Search of a Fundamental Theory.* New York: Oxford University Press.

Chalmers, D. J. (2000). What is a neural correlate of consciousness? In T. Metzinger (Ed.), *Neural Correlates of Consciousness: Empirical and Conceptual Questions* (pp. 17–40). Cambridge, MA: MIT Press.

Cook, M. (2005). *Faces of Science: Portraits.* New York: W.W. Norton.

Collini, E., Wong, C. Y., Wilk, K. E., Curmi, P. M. G., Brumer, P., & Schoes, G. D. (2010). Coherently wired light-harvesting in photosynthetic marine algae at ambient temperature. *Nature, 463,* 644–647.

Crick, F. C. (1995). *The Astonishing Hypothesis: The Scientific Search for the Soul.* New York: Scribner.

Crick, F. C., & Koch, C. (1990). Towards a neurobiological theory of consciousness. *Seminars in Neuroscience, 2,* 263–275.

Crick, F. C., & Koch, C. (1995). Are we aware of neural activity in primary visual cortex? *Nature, 375,* 121–123.

Crick, F. C., & Koch, C. (1998). Consciousness and neuroscience. *Cerebral Cortex, 8,* 97–107.

Crick, F. C., & Koch, C. (2003). A framework for consciousness. *Nature Neuroscience, 6,* 119–126.

Crick, F. C., & Koch, C. (2005). What is the function of the claustrum? *Philosophical Transactions of the Royal Society of London. Series B, Biological Sciences, 360,* 1271–1279.

Dawkins, M. S. (1998). *Through Our Eyes Only? The Search for Animal Consciousness.* New York: Oxford University Press.

Dehaene, S., & Changeux, J.-P. (2011). Experimental and theoretical approaches to conscious processing. *Neuron, 70,* 200–227.

Dehaene, S., Naccache, L., Cohen, L., Le Bihan, D., Mangin, J.-F., Poline, J.-B., et al. (2001). Cerebral mechanisms of word masking and unconscious repetition priming. *Nature Neuroscience, 4,* 752–758.

Desmurget, M., Reilly, K. T., Richard, N., Szathmari, A., Mottolese, C., & Sirigu, A. (2009). Movement intention after parietal cortex stimulations in humans. *Science, 324,* 811–813.

Dower, J. W. (2010). *Cultures of War: Pearl Harbor, Hiroshima, 9–11, Iraq.* New York: W.W. Norton.

Edelman, D. B., & Seth, A. K. (2009). Animal consciousness: A synthetic approach. *Trends in Neurosciences, 32,* 476–484.

Edlund, J.A., Chaumont, N., Hintze, A., Koch, C., Tononi, G., and Adami, C. (2011). Integrated information increases with fitness in the simulated evolution of autonomous agents. *PLoS Computational Biology, 7*(10): e1002236.

Fried, I., Katz, A., McCarthy, G., Sass, K. J., Williamson, P., Spencer, S. S., et al. (1991). Functional organization of human supplementary motor cortex studied by electrical stimulation. *Journal of Neuroscience, 11,* 3656–3666.

Gallant, J. L., Shoup, R. E., & Mazer, J. A. (2000). A human extrastriate area functionally homologous to macaque V4. *Neuron, 27,* 227–235.

Gibson, D. G., & Associates. (2010). Creation of a bacterial cell controlled by a chemically synthesized genome. *Science, 329,* 52–56.

Goldberg, I. I., Harel, M., & Malach, R. (2006). When the brain loses its self: Prefrontal inactivation during sensorimotor processing. *Neuron, 50,* 329–339.

Goodale, M. A., & Milner, A. D. (2004). *Sight Unseen: An Exploration of Conscious and Unconscious Vision.* Oxford, UK: Oxford University Press.

Goodale, M. A., Pélisson, D., & Prablanc, C. (1986). Large adjustments in visually guided reaching do not depend on vision of the hand or perception of target displacement. *Nature, 320,* 748–750.

Gradinaru, V., Mogri, M., Thompson, K. R., Henderson, J. M., & Deisseroth, K. (2009). Optical deconstruction of Parkinsonian neural circuitry. *Science, 324*, 354–359.

Gradinaru, V., Zhang, F., Ramakrishnan, C., Mattis, J., Prakash, R., Diester, I., et al. (2010). Molecular and cellular approaches for diversifying and extending optogenetics. *Cell, 141*, 154–165.

Griffin, D. R. (2001). *Animal Minds: Beyond Cognition to Consciousness.* Chicago, IL: University of Chicago Press.

Haggard, P. (2008). Human volition: Towards a neuroscience of will. *Nature Reviews. Neuroscience, 9*, 934–946.

Hassin, R. R., Uleman, J. S., & Bargh, J. A. (Eds.). (2006). *The New Unconscious.* Oxford, UK: Oxford University Press.

Hasson, U., Nir, Y., Levy, I., Fuhrmann, G., & Malach, R. (2004). Intersubject synchronization of cortical activity during natural vision. *Science, 303*, 1634–1640.

Haynes, J. D., & Rees, G. (2005). Predicting the orientation of invisible stimuli from activity in human primary visual cortex. *Nature Neuroscience, 8*, 686–691.

Herculano-Houzel, S. (2009). The human brain in numbers: A linearly scaled-up primate brain. *Frontiers in Human Neuroscience, 3*, 1–11.

Jeannerod, M. (1997). *The Cognitive Neuroscience of Action.* Oxford, UK: Blackwell.

Jiang, Y., Costello, P., Fang, F., Huang, M., & He, S. (2006). A gender- and sexual orientation-dependent spatial attentional effect of invisible images. *Proceedings of the National Academy of Sciences of the United States of America, 103*, 17048–17052.

Johansson, P., Hall, L., Sikström, S., & Olsson, A. (2005). Failure to detect mismatches between intention and outcome in a simple decision task. *Science, 310*, 116–119.

Jordan, P. (1938). The Verstärkertheorie der Organismen in ihrem gegenwärtigen Stand. *Naturwissenschaften, 33*, 537–545.

Kim, C. Y., & Blake, R. (2005). Psychophysical magic: Rendering the visible invisible. *Trends in Cognitive Sciences, 9*, 381–388.

Koch, C. (2004). *The Quest for Consciousness: A Neurobiological Approach.* Englewood, CO: Roberts & Company.

Koch, C., & Hepp, K. (2011). The relation between quantum mechanics and higher brain functions: Lessons from quantum computation and neurobiology. In R. Y. Chiao, M. L. Cohen, A. J. Leggett, W. D. Phillips, & C. L. Harper, Jr. (Eds.), *Amazing Light: New Light on Physics, Cosmology and Consciousness* (pp. 584–600). New York: Cambridge University Press.

Koch, C., & Segev, I. (2000). Single neurons and their role in information processing. *Nature Neuroscience, 3*, 1171–1177.

Koch, C., & Tononi, G. (2008). Can machines be conscious? *IEEE Spectrum, 45*, 54–59.

Koch, C., & Tononi, G. (2011). A test for consciousness. *Scientific American, 304* (June), 44–47.

Krakauer, J. (1990). *Eiger Dreams.* New York: Lyons & Burford.

Küng, H. (2008). *The Beginning of All Things: Science and Religion.* Cambridge, UK: Wm. B. Eerdmans.

Lafferty, K. D. (2006). Can the common brain parasite, *Toxoplasma gondii*, influence human culture? *Proceedings. Biological Sciences / The Royal Society, 273*, 2749–2755.

Laureys, S. (2005). Death, unconsciousness and the brain. *Nature Reviews. Neuroscience, 6*, 899–909.

Laureys, S., & Tononi, G. (Eds.). (2009). *The Neurology of Consciousness.* New York: Elsevier.

Lein, E. S., & Associates. (2007). Genome-wife atlas of gene expression in the adult mouse brain. *Nature, 445*, 168–176.

Lem, S. (1987). *Peace on Earth.* San Diego: Harcourt.

Lemon, R. N., & Edgley, S. A. (2010). Life without a cerebellum. *Brain, 133,* 652–654.

Leopold, D. A., & Logothetis, N. K. (1999). Multistable phenomena: Changing views in perception. *Trends in Cognitive Sciences, 3,* 254–264.

Libet, B., Gleason, C. A., Wright, E. W., & Pearl, D. K. (1983). Time of conscious intention to act in relation to onset of cerebral activity (readiness-potential). The unconscious initiation of a freely voluntary act. *Brain, 106,* 623–642.

Lin, D., Boyle, M. P., Dollar, P., Lee, H., Lein, E. S., Perona, P., et al. (2011). Functional identification of an aggression locus in the mouse hypothalamus. *Nature, 470,* 221–226.

Logan, G. D., & Crump, M. J. C. (2009). The left hand doesn't know what the right hand is doing: The disruptive effects of attention to the hands in skilled typewriting. *Psychological Science, 20,* 1296–1300.

Logothetis, N. K. (1998). Single units and conscious vision. *Philosophical Transactions of the Royal Society of London. Series B, Biological Sciences, 353,* 1801–1818.

Logothetis, N. K. (2008). What we can do and what we cannot do with fMRI. *Nature, 453,* 869–878.

Macknik, S. L., King, M., Randi, J., Robbins, A., Teller, J. T., & Martinez-Conde, S. (2008). Attention and awareness in stage magic: Turning tricks into research. *Nature Reviews. Neuroscience, 9,* 871–879.

Mann, E. O., & Paulsen, O. (2010). Local field potential oscillations as a cortical soliloquy. *Neuron, 67,* 3–5.

Masland, R. H. (2004) Neuronal cell types. *Current Biology, 14*(13), R497–500.

Massimini, M., Ferrarelli, F., Huber, R., Esser, S. K., Singh, H., & Tononi, G. (2005). Breakdown of cortical effective connectivity during sleep. *Science, 309,* 2228–2232.

Maye, A., Hsieh, C.-H., Sugihara, G., & Brembs, B. (2007). Order in spontaneous behavior. *PLoS ONE, 2,* e443.

McCandliss, B. D., Cohen, L., & Dehaene, S. (2003). The visual word from area: Expertise for reading in the fusiform gyrus. *Trends in Cognitive Sciences, 7,* 293–299.

Monti, M. M., & Associates. (2010). Willful modulation of brain activity in disorders of consciousness. *New England Journal of Medicine, 362,* 579–589.

Mudrik, L., Breska, A., Lamy, D., and Deouell, L. Y. (2011). Integration without awareness: Expanding the limits of unconscious processing. *Psychological Sciences, 22,* 764–770.

Murphy, N., Ellis, G. F., & O'Connor, T. (Eds.). (2009). *Downward Causation and the Neurobiology of Free Will.* Berlin: Springer.

Olby, R. (2009). *Francis Crick: Hunter of Life's Secrets.* New York: Cold Spring Harbor Press.

Owen, A. M., & Associates. (2006). Detecting awareness in the vegetative state. *Science, 313,* 1402.

Parvizi, J., & Damasio, A. R. (2001). Consciousness and the brainstem. *Cognition, 79,* 135–160.

Quian Quiroga, R., Kraskov, A., Koch, C., & Fried, I. (2009). Explicit encoding of multimodal percepts by single neurons in the human brain. *Current Biology, 19,* 1–6.

Quian Quiroga, R., Kreiman, G., Koch, C., & Fried, I. (2008). Sparse but not "Grandmother-cell" coding in the medial temporal lobe. *Trends in Cognitive Science, 12,* 87–89.

Quian Quiroga, R., Reddy, L., Kreiman, G., Koch, C., & Fried, I. (2005). Invariant visual representation by single neurons in the human brain. *Nature, 435,* 1102–1107.

Rauschecker, A. M., Dastjerdi, M., Weiner, K. S., Witthoft, N., Chen, J., Selimbeyoglu, A., & Parvizi, J. (2011). Illusions of visual motion elicited by electrical stimulation of human MT complex. *PLoS ONE 6*(7), e21798.

Ridley, M. (2006). *Francis Crick: Discoverer of the Genetic Code*. New York: HarperCollins.

Rosanova, M., Gosseries, O., Casarotto, S., Boly, M., Casali, A.G., Bruno, M.-A., Mariotti, M., Boveroux, P., Tononi, G., Laureys, S., & Massimini, M. (2012) Recovery of cortical effective connectivity and recovery of consciousness in vegetative patients. *Brain*, in press.

Sacks, O. (2011). *The Mind's Eye*. New York: Knopf.

Schiff, N. D. (2010). Recovery of consciousness after brain injury. In M. S. Gazzaniga (Ed.), *The Cognitive Neurosciences*, 4th ed. (pp. 1123–1136). Cambridge, MA: MIT Press.

Singer, P. (1994). *Rethinking Life and Death: The Collapse of our Traditional Ethics*. New York: St. Martin's Griffin.

Sompolinsky, H. (2005). A scientific perspective on human choice. In Y. Berger & D. Shatz (Eds.), *Judaism, Science, and Moral Responsibility* (pp. 13–44). Lanham, MD: Rowman & Littlefield.

Soon, C. S., Brass, M., Heinze, H.-J., & Haynes, J.-D. (2008). Unconscious determinants of free decisions in the human brain. *Nature Neuroscience, 11*, 543–545.

Squire, L. R. (2009). The legacy of patient H.M. for neuroscience. *Neuron, 61*, 6–9.

Sussman, G. J., & Wisdom, J. (1988). Numerical evidence that the motion of Pluto is chaotic. *Science, 241*, 433–437.

Swanson, L. W. (2012). *Brain Architecture: Understanding the Basic Plan*, 2nd edition. New York: Oxford University Press.

Tononi, G. (2008). Consciousness as integrated information: A provisional manifesto. *Biological Bulletin, 215*, 216–242.

Tononi, G. (2012). *PHI: A Voyage from the Brain to the Soul*. New York: Pantheon Books.

Tononi, G., & Koch, C. (2008). The neural correlates of consciousness: An update. *Annals of the New York Academy of Sciences, 1124*, 239–261.

Tsuchiya, N., & Koch, C. (2005). Continuous flash suppression reduces negative afterimages. *Nature Neuroscience, 8*, 1096–1101.

Turner, M. S. (1999). Large-scale structure from quantum fluctuations in the early universe. *Philosophical Transactions of the Royal Society of London. Series A: Mathematical and Physical Sciences, 357*(1750), 7–20.

Tyndall, J. (1901). *Fragments of Science* (Vol. 2). New York: P.F. Collier and Son.

van Boxtel, J. A., Tsuchiya, N., & Koch, C. (2010). Consciousness and attention: On sufficiency and necessity. *Frontiers in Consciousness Research, 1*, 1–13.

Varki, A. (2009). Human uniqueness and the denial of death. *Nature, 460*, 684.

Vyas, A., Kim, S.-K., Giacomini, N., Boothroyd, J. C., & Sapolsky, R. M. (2007). Behavioral changes induced by *Toxoplasma* infection of rodents are highly specific to aversion of car odors. *Proceedings of the National Academy of Sciences of the United States of America, 104*, 6442–6447.

Watanabe, M., Cheng, K., Murayama, Y., Ueno, K., Asamizuyu, T., Tanaka, K., & Logothetis, N. (2011). Attention but not awareness modulates the BOLD signal in the human V1 during binocular suppression. *Science, 334*, 829–831.

Wegner, D. M. (2003). *The Illusion of Conscious Will*. Cambridge, MA: MIT Press.

Zeki, S. (2001). Localization and globalization in conscious vision. *Annual Review of Neuroscience, 24*, 57–86.

Index

Achromatopsia, 60
Action potential, **16**, 46, 48, 51, 53, 65–66, 100, 103–104, 106, 113, 129, 142–146
Agency, 106–111
Agnosia, 81
Akinetopsia, 62
Allen, Paul, 145
Allen Institute for Brain Science, 145–147
Amygdala, 34, 48, 62, 73, 94, 109
Anderson, David, 142, 176
Angelus Silesius, 165
Animal rights, 140–141
Anthropic principle, 7, 155–156
Artificial intelligence, 25, 131
Attention, selective, 37, 45, 48, **55–57**, 80, 113, 168–169, 176

Baars, Bernie, 122, 170, 173
Basal ganglia, 34, 49, 73, 80, 106, 170
Bias, unconscious, 83–88, 159
Binocular rivalry, 52–54, 168
Biophysics, 16–19, 100–101, 103, 167
Bladerunner, 27, 168
Blind spot, 50
Block, Ned, 5, 168, 173
Bogen, Joseph, 68–69
Braitenberg, Valentino, 15
Butterfly effect, 97

Caenorhabditis elegans, 36, 118, 128
Calculus. *See* Professor Calculus
Capgras delusion, 61
Cerebellum, 42–43, 80, 89, 129, 170, 176
Cerebral cortex, 34–35, 38, 42–43, 47–49, 52–55, 62, 68, 70–74, 82, 89–90, 110, 113, 121–122, 127, 139, 141–142, 145, 168, 170
Changeux, Jean-Pierre, 5, 122, 170, 174
Channelrhodopsin-2 (ChR2), **143–144**, 146, 171
Choice blindness, 87
Claustrum, 34, 73, 113, 163, 171, 174
Coalition. *See* Neuronal coalition

Coherence, 101–103, 170
Coma, 24, 33, 70–71, 139
Compatibilism, 93–95, 111
Complexification, 7, 133
Computer consciousness. *See* Consciousness, computer
Chaos, 97–98, 101, 111
Complexity, 7, 36, 63, 117, 119, 133–134, 141, 147, 165
Cell types, 16, 43, 54, 141–142, 146–147, 171, 176
California Institute of Technology (Caltech), 6, 17, 20, 32, 66, 69, 142, 147, 159
Chalmers, David, 3, 5, 26, 124, 167, 168, 170, 173–174
Computer programming, 15
Concept neurons, **65–67**, 100, 169
Consciousness
animal, 34–36, 53–54, 116, 145–147
computer, 2, 131, 170
epiphenomenal, 31, 129
different from attention, **55–57**, 168–169, 176–177
definition, 32–34, 42
dream, 33, 44, 50–52, 70, 138
feedback loop for, 43, 48, 53–54, 89, 114, 146
function, **29–31**
-meter, 7, 72, 131, 138–140
self-, 36–39
visual, 48–53, 71, 89, 125, 146
Continuous flash suppression (CFS), **45–46**, 48, 52, 55, 168, 177
Corpus callosum, **68**–70, 126, 142
Cortex. *See* Cerebral cortex
Cortico-thalamic complex, **34**, 51, 70, 73, 76, 89–90, 111, 124, 140, 146
Crick, Francis, 5, 19–21, 30–31, 42–44, 48–49, 53, 68, 78, 90, 113–114, 121–124, 137, 146, 153–154, 162–163, 167–171, 174

Dante, Alighieri, 8–9, 11, 164
Decision making, 30, 43, 49, 85–88, 94, 101,
    105–107, 111, 169
Deep sleep, 24, 33, 52, 70, 126, 132,
    138–139
Dehaene, Stanislas, 5, 122, 147, 168,
    170–171, 174, 176
Der Ring des Nibelungen, 15, 94–95
Descartes, René, 23–24, 68, 92, 102, 114,
    123, 134, 150–151, 165, 168
Determinism, 93, 95–96, 98, 100–102, 110
Deterministic chaos. See Chaos
DNA, 5, 24, 89, 117, 145
Dogs, 35, 72, 90, 115–116, 120, 140, 151,
    160, 162
Dostoyevsky, Fyodor, 27
Dual-aspect theory, 124
Dualism, 23, 149–152

Eccles, John, 103–104, 151
Emergence, 7, 116–119
Emotions, 38–39, 44, 61, 77, 115
Enabling factor, 73
Entanglement, 102–103, 105, 114
Enteric nervous system, 28–29
Epicurus, 164
Essential node, 62
Electroencephalography (EEG), 18, 40, 54,
    64, 71, 126, **138–139**, 170
Evolution, 2, 4, 7, 11, 19, 31, 36, 38, 101,
    115, 129, 133–134, 145, 153, 169, 174

Face blindness, **60–61**
Faith, 11–12, 149, 153, 156, 159, 166
Feedback loop, 43, 48, 53–54, 89, 114,
    146
First-person perspective, 24
Fried, Itzhak, 64–67, 110, 170, 174, 176
Freud, Sigmund, 6, 76–77
Functionalism, 120–121
Functional magnetic resonance imaging
    (fMRI), **47**, 72, 142, 147, 168

Global workspace theory, 122, 170
Grandmother neurons, 169

Hard Problem, **3**, 5, 26, 137, 167
Huxley, Thomas 31, 168
Hypothalamus, 73, 144

Indeterminism, 101–102
Inferior temporal cortex, 53–54, 82
Information theory, 115, 123–124, 170
Inhibitory interneuron, 142, 144, 146
Integrated information theory (IIT), 6, 60,
    124–131, 146, 152–153, 170, 177
Intralaminar thalamic nuclei, 73

Jennifer Aniston neurons. See Concept
    neurons

Laureys, Steven, 5, 139, 167, 169, 175
Leibniz, Gottfried, 26, 119, 131, 133
Lesion, 43
Libertarian free will, 92, 102, 104, 151
Libet, Benjamin, 105–107, 170, 176
Logothetis, Nikos, 53–54, 168, 176–177
Lucretius, 102

Macaque monkey. See Monkey
Magnetic resonance imaging (MRI), 6, 45,
    47, 60, 168
Masking, 46, 52, 55, 147, 168
Massachusetts Institute of Technology
    (MIT), 17, 19
Matrix, The, 44, 91–92, 111
Medial temporal lobe, **64–67**, 169
Mind–body dualism. See Dualism
MindScope, Project, 145
Minimal conscious state (MCS), 71–72,
    139, 170
Mirror test, 36–37
Monkey, 32, 35, 37, 49, 51, 53, 57, 82–83, 102,
    141, 145
Monad, 26, 119, 125–126
Monism, 15
Monod, Jacques, 4, 5
Mouse, 33, 35, 57, 141, 144–147, 160, 171
Mycoplasma mycoides, 24

Neocortex, **34**, 52, 122, 142, 145
Neuronal coalition, 16, 46, 54, 63, 66, 71,
    73–74, 89, 90, 117, 146–147
Neuronal (neural) correlates of
    consciousness, **42–44**, 49, 51, 53, 67,
    89–90, 114, 125, 128, 168, 174, 177
Newton, Isaac, 95–98, 155
Nietzsche, Friedrich, 4, 15, 75, 77, 150
Non-rapid-eye-movement sleep. See Deep
    sleep

Optogenetics, 26, **143–144**, 171

Panpsychism, 131–134
Persistent vegetative state (PVS), 71–72,
    170
Pharmacogenetics, 144
Poggio, Tomaso, 15
Popper, Karl, 103–104, 151
Prefrontal cortex, 43–44, 49, 54, 73, 80, 82,
    89, 94, 121–122, 140
Primary visual cortex, 47–55, 60, 82, 89,
    146, 168, 174–175
Priming, 85, 169
Professor, being a, 14, 17–18

Professor Calculus, 14
Property dualism, 152
Prosopagnosia. *See* Face blindness
Pyramidal neuron, 44, 49, 68, 89, 106, 113, 122, 129, 142

Qualia, **27–29,** 30–32, 43, 54, 111, 121, 128, 130–131
Quantum mechanics, 98–104, 110–111, 114, 154, 169, 170, 177

Rapid-eye-movement (REM) sleep, 52, 138
Readiness potential, 105–107, 176
Rees, Geraint, 5, 51, 168, 175
Religion, 3–4, 12, 86, 116, 153, 159, 165, 171
Retina, 46, 48–51, 53, 82, 89–90, 142, 144
RNA, 89, 134
Romantic reductionist, 8

Saccade, 50, 79
Sacks, Oliver, 20, 61, 169
Selective visual attention. *See* Attention, selective
Self-consciousness. *See* Consciousness, self
Sherlock Holmes, 41, 68, 128, 158
Silver Blaze, 128
Singer, Peter, 160, 171, 177
Singer, Wolf, 5
Soul, 3–4, 23, 93–94, 103, 111, 116, 149–152
Sparse representation, 66
Sperry, Roger, 69–70
Spike. *See* Action potential
Split brain patients, **69–70,** 126, 170
Spotlight of attention. *See* Attention, selective
Strange loop, 38
Synchronization. *See* Synchrony
Synchrony, 18, 48, 53, 73, 125, 129, 167

Teilhard de Chardin, Pierre, 7, 133–134
Thalamus, 34, 49, 70, 73, 142, 170
Theory of integration information. *See* Integrated information theory
Thermal motion, 98, 100
Third-person perspective, 26
Tiling, 142
Tintin, 14, 68
Tononi, Giulio, 5, 6, 124–126, 128, 130–131, 134, 138–140, 167–168, 170, 174–177
Toxoplasmosis, 108–109
Transcranial magnetic stimulation (TMS), 138–140, 170

Unconscious. *See* Unconscious processing
Unconscious processing, 30, 47–48, 61, 72, 76–90, 111–112, 119, 129–130, 161, 163, 168–169, 175

V1. *See* Primary visual cortex
Vegetarianism, 160
Vegetative state (VS), 34, 52, 71–72, 139–140, 169–170

Wagner, Richard, 14–15, 94, 127
Watson, James, 5, 20
Wilde, Oscar, 2

Zombie agents or actions, 30–31, 33, **78–81,** 89–90, 108, 122, 130, 146